Lecture Notes in Computer Science 6348

Commenced Publication in 1973
Founding and Former Series Editors:
Gerhard Goos, Juris Hartmanis, and Jan van Leeuwen

Editorial Board

Alan Dearle   Roberto V. Zicari (Eds.)

Objects and Databases

Third International Conference, ICOODB 2010
Frankfurt/Main, Germany, September 28-30, 2010
Proceedings

 Springer

Volume Editors

Alan Dearle
University of St Andrews, School of Computer Science
Jack Cole Building, North Haugh, St Andrews, KY16 9SX, UK
E-mail: al@cs.st-andrews.ac.uk

Roberto V. Zicari
Johann-Wolfgang-Goethe University, Institute of Computer Science
Robert-Mayer-Str. 10, 60325 Frankfurt/Main, Germany
E-mail: roberto@zicari.de

Library of Congress Control Number: 2010934756

CR Subject Classification (1998): H.2, H.4, D.2, H.3, H.5, C.2.4

LNCS Sublibrary: SL 3 – Information Systems and Application, incl. Internet/Web and HCI

ISSN	0302-9743
ISBN-10	3-642-16091-3 Springer Berlin Heidelberg New York
ISBN-13	978-3-642-16091-2 Springer Berlin Heidelberg New York

springer.com

© Springer-Verlag Berlin Heidelberg 2010
Printed in Germany

Typesetting: Camera-ready by author, data conversion by Scientific Publishing Services, Chennai, India
Printed on acid-free paper   06/3180

Preface

According to Francois Bancillon and Won Kim [SIGMOD RECORD, Vol. 19, No. 4, December 1990], object-oriented databases started in around 1983. Twenty-seven years later this publication contains the proceedings of the Third International Conference on Object-Oriented Databases (ICOODB 2010).

Two questions arise from this – why only the third, and what is of interest in the field of object-oriented databases in 2010? The first question is easy – in the 1980s and 1990s there were a number of conferences supporting the community – the International Workshops on Persistent Object Systems started by Malcolm Atkinson and Ron Morrison, the EDBT series, and the International Workshop on Database Programming Languages. These database-oriented conferences complimented other OO conferences including OOPSLA and ECOOP, but towards the end of the last century they dwindled in popularity and eventually died out.

In 2008 the First International Conference on Object Databases was held in Berlin. In 2009 the second ICOODB conference was held at the ETH in Zurich as a scientific peer-reviewed conference.

What is particular about ICOODB is that the conference series was established to address the needs of both industry and researchers who had an interest in object databases, in innovative ways to bring objects and databases together and in alternatives/extensions to relational databases. The first conference set the mould for those to follow – a combination of theory and practice with one day focusing on the theory of object databases and the second focusing on their practical use and implementation.

The second conference helped re-establish the research agenda in the field of objects and databases.

This third conference took place in Frankfurt, and continued the tradition set forth in Zurich as a scientific peer-reviewed conference. The key goal of the 2010 conference remained to bring together developers, users, and researchers in objects and database technologies.

So what of the second question – what is of interest in the field of object-oriented databases in 2010? This is more complex. Firstly, the management of large bodies of programs and data that are potentially long-lived is of the upmost social and economic importance. We have an increased and increasing need for technologies that support the design, construction, maintenance, and operation of long-lived, concurrently accessed, and potentially large bodies of data and programs. A recent (2007) quote from Microsoft illustrates the current technological situation well:

"Most programs written today manipulate data in one way or another and often this data is stored in a relational database. Yet there is a huge divide between modern programming languages and databases in how they represent and

manipulate information. This impedance mismatch is visible in multiple ways. Most notable is that programming languages access information in databases through APIs that require queries to be specified as text strings. These queries are significant portions of the program logic. Yet they are opaque to the language, unable to benefit from compile-time verification and design-time features like IntelliSense." [Kulkarni, D., Bolognese, L., Warren, M., Hejlsberg, A., George, K.: LINQ to SQL: .NET Language-Integrated Query for Relational Data. http://msdn.micro soft.com/en-gb/library/bb425822.aspx (2007)]

Secondly, the world has changed since 1983 (pre-Web) and database applications are pretty much the norm in the Web space of 2010. Most of these applications are being coded in object-oriented languages, primarily Java but also in .Net languages such as C# and hybrid languages such as Python and Ruby. These Web-based database applications have complex data management needs. As stated on the ICOODB 2008 Web pages, *"Object databases are the right choice for a certain class of application, to save developers cost and time and help them to build more feature rich OO applications"*. Such applications are a new driver for the development of OODB systems.

Another 21st century driver is mobile application systems. Smart devices such as iPhones are now ubiquitous. These all have complex database requirements – both for local (mobile) databases and those associated with Web servers; highlighting the variations in scale that database systems must address – from small databases to support personal data up to large scientific databases. We require appropriate technologies to address both these domains and all those in between.

In the 1980s the data-models such as hierarchical and relational were characterized by having a strong separation between schema and data. Since then new data-models have arrived with a much looser relationship between schema and data. Many of the persistent and object-oriented systems of the last three decades have featured such a loose relationship. This trend has been continued in semi-structured databases, which have recently become prevalent, fuelled by the Web, XML, and the need to model and compute over richer data sets.

This need has also fuelled another trend – the so-called NoSQL model that in turn is driven by another trend namely cloud computing. Like semi-structured databases, the NoSQL data models move away from the traditional relational data model favouring instead horizontally partitioned collections of records. NoSQL databases typically do not provide an SQL query language and instead rely upon the ability to perform massively parallel computation against the data.

Lastly, the use of hybrid object-relational technologies remains strong. Typically data represented in object-oriented language is mapped to and from a relational schema. Such Object-Relational Mapping (ORM) technologies are an important feature of the 21st Century Web Service provision.

The ICOODB 2010 program committee recognized that the world of data management is changing. We therefore expanded the focus of ICOODB 2010 to include new areas that are becoming hot topics both in academia and industry, such as the linkage to service platforms, operation within scalable (cloud) platforms, object-relational bindings, NoSQL databases, and new approaches to concurrency control.

These recent trends in what might be termed alternative database technologies are reflected in the papers that we have selected for the third ICOODB.

Three of the papers relate to ORM. One, "Solving ORM by MAGIC: MApping GeneratIon and Composition," is related to the problem of maintaining the mappings between objects and relations – a problem that has been called the Vietnam of Computer Science [The Vietnam of Computer Science, Ted Neward – June 26, 2006 ODBMS.ORG (www.odbms.org)]. The second, "Closing Schemas in Object-Relational Databases," is concerned with schema closure – that is ensuring that the types used in a particular domain are complete in that the objects involved in a computation do not contain references to types that are not defined outside that domain. The last, "A Comperative Study of the Features and Performance of ORM Tools in a .NET Environment," combines two areas – a comparative study of the ORM techniques used within the .Net environment.

XML semi-structured data models are addressed in one paper entitled "Object-Oriented Constraints for XML Schema." This paper examines the type system of XML schema and proposes an object-oriented assertion language that is capable of expressing concepts such as range constraints, keys, and referential integrity that are not normally expressible in a programming language. This paper addresses the problem domain of expressing object-XML rather than object-relational mappings.

The mapping of data models onto Web-based systems is addressed in the paper "Data Model Driven Implementation of Web Cooperation Systems with Tricia." The paper demonstrates the benefits, namely expressiveness, modularity, and reuse, derived from the use of a data modeling framework by application developers. The development cycle of data-intensive application systems is also examined in the paper entitled "Revisiting Schema Evolution in Object Databases in Support of Agile Development."

Querying and data models remain another mainstay of the database world and the proceedings contain two papers on that topic. One of these, "A Flexible Object Model and Algebra for Uniform Access to Object Databases," is concerned with query optimization in embedded language contexts. The other, "Query Optimization by Result Caching in the Stack-Based Approach," relates to query algebras for object databases.

Scientific database applications in the domains of genomic, multimedia, and geo-spatial data have requirements for handling complex binary data objects that are highly structured, large, and of variable length. This domain is the subject of "iBLOB: Complex Object Management in Databases through Intelligent Binary Large Objects" which combines the domains of application programming and type systems and proposes both a new conceptual framework and a novel data type.

The paper "The Case for Object Databases in Cloud Data Management" is very forward-looking. The author argues that there are strong indicators that *the full potential of cloud computing data management can only be leveraged by exploiting object database technologies.* The paper examines the challenges of cloud computing data management and shows the opportunities for object database technologies. Perhaps this will be the start of a whole new research domain unanticipated by the early object database pioneers such as Bancillon, Kim, and Atkinson in the 1980s.

July 2010

Alan Dearle
Roberto V. Zicari

Organization

Steering Committee

Mike Card	Syracuse Research, USA
Rick Cattell	Independent Consultant, USA
William R. Cook	University of Texas at Austin, USA
Stefan Edlich	TFH Berlin, Germany
Anat Gafni	db4objects, USA
Robert Greene	Versant Corporation, USA
Leon Guzenda	Objectivity Inc., USA
Moira C. Norrie	ETH Zurich, Switzerland
James Paterson	Glasgow Caledonian University, UK
Roberto V. Zicari	Goethe University Frankfurt, Germany

Conference Organization

ICOODB 2010 was organized by the database group (DBIS) at the Institute of Computer Science, Goethe University Frankfurt in cooperation with ODBMS.ORG (www.odbms.org).

General Chair

Roberto V. Zicari	Goethe University Frankfurt, Germany

Scientific Program Chairs

Alan Dearle	University of St Andrews, Scotland
Roberto V. Zicari	Goethe University Frankfurt, Germany

Industrial Track Chairs

Anat Gafni	db4objects, USA
Roberto V. Zicari	Goethe University Frankfurt, Germany

Tutorial Track Chairs

Beat Signer	Vrije Universiteit Brussel, Belgium
Jim Paterson	Glasgow Caledonian University, UK

Workshops Chairs

Stefan Edlich   TFH Berlin, Germany
Jim Paterson   Glasgow Caledonian University, UK

Demonstrations and Posters Chairs

Stefan Edlich   TFH Berlin, Germany
Roberto V. Zicari   Goethe University Frankfurt, Germany

Local Organization

Natascha Hoebel   Goethe University Frankfurt, Germany
Naveed Mushtaq   Goethe University Frankfurt, Germany
Clemens Schefels   Goethe University Frankfurt, Germany
Marion Terrell   Goethe University Frankfurt, Germany
Karsten Tolle   Goethe University Frankfurt, Germany

Program Committee

Suad Alagic   University of Southern Maine, USA
William R. Cook   University of Texas at Austin, USA
Suzanne W. Dietrich   Arizona State University, USA
Manfred Jeuseld   Tilburg University, The Netherlands
David Jordan   SAS Institute, Inc., Germany
Michael Grossniklaus   Politecnico di Milano, Italy
Giovanna Guerrini   University of Genoa, Italy
Theo Härder   TU Kaiserslautern, Germany
Natascha Hoebel   Goethe University Frankfurt, Germany
Daniel Lieuwen   Google Inc., USA
Moira C. Norrie   ETH Zurich, Switzerland
Tore J.M. Risch   University of Uppsala, Sweden
Elke A. Rundensteiner   Worcester Polytechnic Institute, USA
Clemens Schefels   Goethe University Frankfurt, Germany
Nicolas Spyratos   University of Paris South, France
Kazimierz Subieta   Polish-Japanese Institute of Tech., Poland
Karsten Tolle   Goethe University Frankfurt, Germany
Susan D. Urban   Arizona State University, USA

Additional Referees

Sebastian Bächle   Dimitris Kotzinos   Karsten Schmidt
Véronique Benzaken   Viet Phan-Luong   Andreas Weiner
Vassilis Christophides

Sponsoring Institutions

Versant Corporation
db4objects
InterSystems Corporation
IBM Deutschland
sones

Supported by

DBIS - Institute of Informatics, Goethe University Frankfurt
Goethe University Frankfurt
ODBMS.ORG

Media Partners

CRC Press Taylor and Francis
OBJEKTspektrum
Apress

Table of Contents

Keynotes

Search Computing Challenges and Directions 1
   Stefano Ceri, Daniele Braga, Francesco Corcoglioniti,
   Michael Grossniklaus, and Salvatore Vadacca

Searching the Web of Objects 6
   Ricardo Baeza-Yates

Unifying Remote Data, Remote Procedures, and Web Services 8
   William R. Cook

Keynote Panel: "New and Old Data Stores" (Abstract) 9
   Ulf Michael Widenius, Michael Keith, Patrick Linskey,
   Robert Greene, Leon Guzenda, and Peter Neubauer

Regular Papers

Revisiting Schema Evolution in Object Databases in Support of Agile
Development 10
   Tilmann Zäschke and Moira C. Norrie

The Case for Object Databases in Cloud Data Management 25
   Michael Grossniklaus

Query Optimization by Result Caching in the Stack-Based Approach ... 40
   Piotr Cybula and Kazimierz Subieta

A Flexible Object Model and Algebra for Uniform Access to Object
Databases.. 55
   Michael Grossniklaus, Alexandre de Spindler,
   Christoph Zimmerli, and Moira C. Norrie

Data Model Driven Implementation of Web Cooperation Systems with
Tricia 70
   Thomas Büchner, Florian Matthes, and Christian Neubert

iBLOB: Complex Object Management in Databases through Intelligent
Binary Large Objects 85
   Tao Chen, Arif Khan, Markus Schneider, and Ganesh Viswanathan

Object-Oriented Constraints for XML Schema 100
   Suad Alagić, Philip A. Bernstein, and Ruchi Jairath

Solving ORM by MAGIC: MApping GeneratIon and Composition   118
    David Kensche, Christoph Quix, Xiang Li, and Sandra Geisler

Closing Schemas in Object-Relational Databases   133
    Manuel Torres, José Samos, and Eladio Garví

A Comparative Study of the Features and Performance of ORM Tools
in a .NET Environment   147
    Stevica Cvetković and Dragan Janković

Author Index   159

Search Computing Challenges and Directions

Stefano Ceri, Daniele Braga, Francesco Corcoglioniti,
Michael Grossniklaus, and Salvatore Vadacca

Dipartimento di Elettronica e Informazione, Politecnico di Milano
P.za Leonardo Da Vinci, 32
I-20133 Milano, Italy
{ceri,braga,corcoglioniti,grossniklaus,vadacca}@elet.polimi.it

Abstract. Search Computing (SeCo)[1] is a project funded by the European Research Council (ERC). It focuses on building the answers to complex search queries like "Where can I attend an interesting conference in my field close to a sunny beach?" by interacting with a constellation of cooperating search services, using ranking and joining of results as the dominant factors for service composition. SeCo started on November 2008 and will last 5 years. This paper will give a general introduction to the Search Computing approach and then focus on its query optimization and execution engine, the aspect of the project which is most tightly related to "objects and databases" technologies.

1 Introduction

Search engine technology provides worldwide users with the ability to get to the "best" Internet pages with the simplest possible query language. However, this simple query paradigm shows its limits when search is complex and the query cannot be compressed to keywords, or the query results are complex and cannot be included into a single page.

Performing a complex search process with a conventional search engine challenges the user's ability to break the process into several tasks, then interacting with the search engine multiple times, and then mentally reconstructing a global solution. Normally, each task can be made small enough to address a single domain. However, the answer of the global process is usually based upon comparisons and trade-offs which span over the various domains of interest, and require compositional activities performed in the user's mind (maybe augmented with notes). Such processes take place routinely, but they are far from being supported by current technology.

We define *search computing systems* [1] as a new class of systems aimed at responding to multi-domain queries, i.e., queries over multiple semantic fields of interest. Such systems support users in expressing complex queries, then decomposing queries into subqueries that can be addressed to a specific data source, then assembling complete results from partial answers, and making sure that the

---

[1] http://www.search-computing.eu

A. Dearle and R.V. Zicari (Eds.): ICOODB 2010, LNCS 6348, pp. 1–5, 2010.

order in which complete results are produced takes into account their "global" ranking. In addition to generic search engines, this process may also involve more specific data sources and search systems. In accordance with the current software trends, we assume that component systems will be accessible through service interfaces, possibly hiding data management languages.

Building such systems requires solving many research problems. First, data sources must be semantically described so as to enable a query understanding and decomposition process. Then, a rank-aware query execution technology should support scalable and efficient query execution execution. Then, query results should be made available to users, in formats which allow their browsing and their comparative visualization. All these research areas are present in the SeCo project, together with many other research focuses, describing interaction aspects, rank-join theory, application design methods and scenarios, design tools, business and legal models.

This paper focuses on the query optimization and execution engine, which exhibits many interesting properties from a systems' perspective.

2   Query Optimization and Execution Engine

In this section, we describe Panta Rhei, the physical query algebra and runtime used in the SeCo query processor. We start by introducing the underlying data and control models, and then describe the topology of query execution plans, that are graphs consisting of nodes that represent operands (units) and edges that represent the data and control flow. We then define all types of edges and units in detail. Finally, we describe how query plans can be composed by giving a minimal set of recursive rewriting rules to define the concept of a *well-formed* plan. An overview of the concepts of Panta Rhei is given in Figure 1.

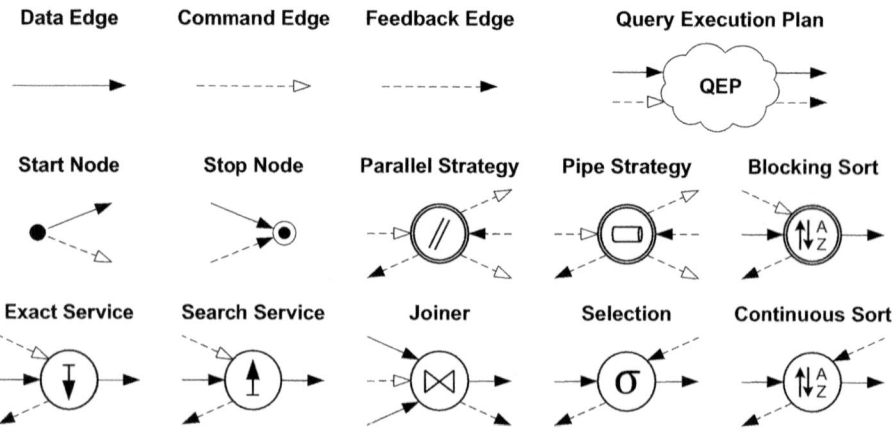

Fig. 1. Panta Rhei concepts

2.1   Data and Control Flow

The *data model* is based on an abstraction that represents the underlying data sources in terms of access patterns that define a list of input, output and ranked attributes which can be atomic or set-valued. Result tuples are progressively composed by using service results as the query evaluation progresses. The *data flow* of a query execution plan consists of **data edges** that form a directed acyclic graph. Every data edge carries tuples whose schema is obtained as the concatenation of all the schemes of the services which are invoked by antecedent nodes of that edge.

The *control model* of the execution engine addresses the fact that, in Search Computing, query processing involves a wide scope of data sources, ranging from traditional databases to Web services and search engines. If the query planner and optimizer can rely on accurate data statistics and estimations of the behavior of the data sources involved in a query, it is possible to completely specify the execution of a query at compile-time. However, if this information is not available at compile-time, the control model must be flexible enough to adapt at run-time. Moreover, plans which want to guarantee optimality (top-k) must adapt their behavior to the actual ranking values which are read from service results. The *control flow* of a query execution plan is bidirectional and comprises **command edges** and **feedback edges** to support both forward and backward scheduling of plans. The forward control flow transports instructions to a query execution plan indicating how the tuples in input must be considered by the plan, the backward control flow reports as feedback statistical data characterizing the plan execution.

2.2   Query Execution Plans

A **query execution plan** (QEP) is a component that accepts an incoming data and control flow edge and produces a data and control flow edge in output. The incoming data edge transmits chunks of tuples in input to the QEP, while the outgoing data edge transmits chunks of tuples to a downstream query execution plan or to the stop node. The incoming control edge expresses how the tuples in input should be processed within the QEP. The outgoing control edge transmits feedback data about the execution of the QEP.

The **start node** injects the constant values specified by the query into the query execution plan along the data flow edge. Additionally, it transmits the "start command" along the control flow edge. Apart from acting as a sink for all feedbacks, the **stop node** collects the results of a query execution plan and makes them available to clients of the execution engine. There is exactly one start and one end node per query.

The **parallel** and **pipe strategy unit** control two query execution plans that perform a parallel or pipe join, respectively, as illustrated in Figure 2. A parallel execution plan is built by two QEPs which are invoked in parallel, followed by a **joiner unit** which receives chunks of tuples from two different QEPs and joins them as instructed by the parallel strategy unit. In contrast, a pipe execution plan is built by two QEPs which are invoked in sequence. The join of results is implicitly performed by the second QEP, whose input data flow is produced by the first QEP. In both cases, the actual strategy of the strategy unit depends on whether the query execution plan is scheduled in forward or backward manner. Forward strategies are static, completely pre-configured by an optimizer. Backward strategies are dynamic and internally generated or altered.

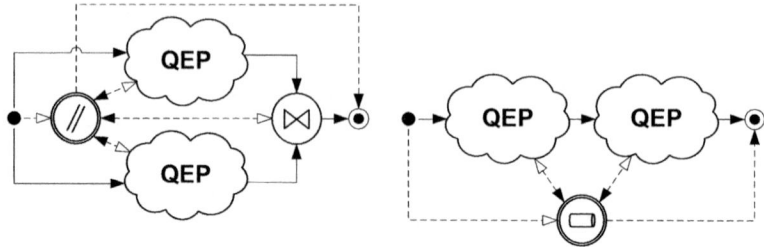

Fig. 2. QEPs for pipe and parallel joins

In our model, data sources are classified in two neatly distinguished categories, called *search* and *exact services*. While this classification is of course a simplification, it is capable of capturing the characteristics of most actual services. Accordingly, the **search** and **exact service units** invoke services of the respective types, using the given input to return result tuples. Search services exhibit a behavior similar to Web search engines: results are unbound, ranked and chunked, and normally there is no need to obtain a complete result, but only the first chunks. Exact services produce a finite set of tuples that represent the exact (and thus complete) response to the service call with the given the input parameters. The output tuples are neither ranked nor chunked.

The **blocking sort unit** buffers all chunk flowing along its input data edge until it receives the EOF message. At that point, it sorts the tuples across all chunks according to a given sort function and emits a sorted output, structured as a series of new chunks of a configurable size. In contrast, the **continuous sort unit** sorts data on a per-chunk basis as the data flows along a data edge. Each chunk in input corresponds to one sorted chunk in output, containing the same tuples. The **selection unit** filters chunks of tuples according to a configured selection predicate. Since the selection unit does not re-chunk the tuples, the chunk size can decrease due to the selectivity of the given predicate.

2.3   Plan Composition

The following set of production rules defines how QEPs can be recursively substituted to compose more complex plans. The axiom QEP consists of a single service unit with a start node as predecessor and a stop node as successor. Rules indicate that the pipe or parallel composition of two QEPs gives a QEP, and that a QEP can be composed with any unary units (i.e. units with a single input and output data and control flow) yielding a QEP. Plans obtained by arbitrary applications of these rules to the axiom are called well-formed and have associated well-defined semantics.

$$QEP := \textcircled{\uparrow} \vee \textcircled{\downarrow} \qquad\qquad QEP := QEP \bowtie_{pipe} QEP$$
$$QEP := QEP \cup \{ \textcircled{$\uparrow_2^A$}, \textcircled{$\uparrow_2^A$}, \textcircled{σ} \} \qquad QEP := QEP \bowtie_{parallel} QEP$$

2.4   Example

The example query shown in Fig. 3 searches for a good and recent adventure movie in a theater not too far from the user's home and a good restaurant nearby.

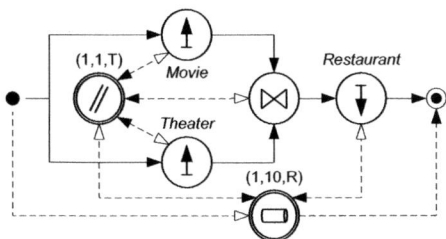

Fig. 3. A simple example query

The query execution plan uses a pipe join strategy unit to schedule the outer join. The first sub-plan of the outer pipe join is the parallel join of the movie and theater services. The results of these fetches are combined by the joiner unit and forwarded to the second sub-plan of the outer pipe join, i.e., the restaurant service.

3   Conclusion

Panta Rhei is the core component of a complete architecture for building search computing systems that process queries over data sources ranging from traditional databases to Web services and search engines. Our engine effectively optimizes the combination of data sources in two ways, namely through pipe and parallel joins. The controllers of these operations implement several strategies that guarantee good performance by limiting the performance bottleneck of service calls. In future work, we plan to port the current implementation to a tuple space environment in order to obtain a scalable Panta Rhei implementation for cloud computing.

References

1. Ceri, S., Brambilla, M. (eds.): Search Computing: Challenges and Directions. Springer, Heidelberg (March 2010)

Searching the Web of Objects

Ricardo Baeza-Yates

Yahoo! Research, Barcelona, Spain

Abstract. We present a pragmatic approach to search the Web of Objects, that is, a Web where entities such as people or places are recognized and exploited. We outline a search architecture where information extraction and semantic technologies play key roles. This architecture has to cope with incompleteness as well as noise to expand the capabilities of current search engines. The main open problems for research are related with recognizing the entities in the query and ranking objects. We show some of these ideas through features or demos already available.

1 Summary

The Web of Objects (WOO) is a new way of organizing Web content in terms of entities and relationships between them. That is, a Web page is now a set of objects and relations among them. The WOO is related to the Semantic Web initiative and the Open Linking Data project[1] is one of the best examples of what could be this Web in the future. Important basic objects are entities like persons, places, dates, etc. This is also called the Web of Concepts [7]. Notice that the WOO is different from the Web of Things, which are physical objects that contain embedded devices connected to Internet that are integrated through the Web.

Searching the WOO implies several challenges, at pre-processing time and at query time, as we outline in [4]. At pre-processing time the main ones are:

1. Object crawling.
2. Object extraction [6,8].
3. Object disambiguation as well as object reconciliation.
4. Object normalization.

At query time we need to:

1. Extract objects from the query and the context of the search.
2. Predict the intent of the query [1,9].
3. Ranking the objects matching the query [13].
4. Laying out the answer (not trivial when we have multiple object types).

---

[1] URL:
http://esw.w3.org/topic/SweoIG/TaskForces/
CommunityProjects/LinkingOpenData

A. Dearle and R.V. Zicari (Eds.): ICOODB 2010, LNCS 6348, pp. 6–7, 2010.
© Springer-Verlag Berlin Heidelberg 2010

To these problems we have to add horizontal ones like scalability, on-line performance and the integration of the social dimension [11]. More details on these challenges can be found in [5].

Important tools to solve these problems include machine learning applied to information extraction (e.g. see [10]), semantic Web technology [3] and Web usage mining [2,12].

References

1. Baeza-Yates, R., Calderón-Benavides, L., González-Caro, C.: The intention behind Web queries. In: Crestani, F., Ferragina, P., Sanderson, M. (eds.) SPIRE 2006. LNCS, vol. 4209, pp. 98–109. Springer, Heidelberg (2006)
2. Baeza-Yates, R., Tiberi, A.: Extracting Semantic Relations from Query Logs. In: ACM KDD 2007, San Jose, California, USA, pp. 76–85 (August 2007)
3. Baeza-Yates, R., Mika, P., Zaragoza, H.: Search, Web 2.0, and the Semantic Web. In Trends and Controversies: Near-Term Prospects for Semantic Technologies. In: Benjamins, R. (ed.) IEEE Intelligent Systems, vol. 23 (1), pp. 80–82 (2008)
4. Baeza-Yates, R., Ciaramita, M., Mika, P., Zaragoza, H.: Towards Semantic Search. In: Kapetanios, E., Sugumaran, V., Spiliopoulou, M. (eds.) NLDB 2008. LNCS, vol. 5039, pp. 4–11. Springer, Heidelberg (2008)
5. Baeza-Yates, R., Raghavan, P.: Next Generation Web Search. In: Ceri, S., Brambilla, M. (eds.) Search Computing. LNCS, vol. 5950, pp. 11–23. Springer, Heidelberg (2010)
6. Chen, F., Doan, A., Yang, J., Ramakrishnan, R.: Efficient Information Extraction over Evolving Text Data. In: ICDE, pp. 943–952 (2008)
7. Dalvi, N., Kumar, R., Pang, B., Ramakrishnan, R., Tomkins, A., Bohannon, P., Keerthi, S., Merugu, S.: A Web of concepts. In: PODS, pp. 1–12 (2009)
8. Doan, A., Naughton, J., Ramakrishnan, R., Baid, A., Chai, X., Chen, F., Chen, T., Chu, E., DeRose, P., Gao, B., Gokhale, C., Huang, J., Shen, W., Vuong, B.-Q.: Information extraction challenges in managing unstructured data. SIGMOD Record 37(4), 14–20 (2008)
9. Jansen, B.J., Booth, D.L., Spink, A.: Determining the user intent of Web search engine queries. In: Proc. of the 16th international conference on World Wide Web, pp. 1149–1150. ACM Press, New York (2007)
10. Mika, P., Ciaramita, M., Zaragoza, H., Atserias, J.: Learning to Tag and Tagging to Learn: A Case Study on Wikipedia. IEEE Intelligent Systems 23(5), 27–33 (2008)
11. Ramakrishnan, R., Tomkins, A.: Toward a PeopleWeb. Computer 40(8), 63–72 (2007)
12. Surowiecki, J.: The Wisdom of Crowds. Random House, New York (2004)
13. Zaragoza, H., Rode, H., Mika, P., Atserias, J., Ciaramita, M., Attardi, G.: Ranking Very Many Typed Entities on Wikipedia. In: CIKM 2007: Proceedings of the sixteenth ACM international conference on Information and Knowledge Management, Lisbon, Portugal (2007)

Unifying Remote Data,
Remote Procedures,
and Web Services

William R. Cook

Department of Computer Science, University of Texas at Austin

Abstract. Most large-scale applications integrate remote services and/
or transactional databases. Yet building software that efficiently invokes
distributed service or accesses relational databases is still quite difficult.
Existing approaches to these problems are based on the Remote Procedure Call (RPC), Object-Relational Mapping (ORM), or Web Services
(WS). RPCs have been generalized to support distributed object systems. ORM tools generally support a form of query sublanguage for efficient object selection, but it is not well-integrated with the host language.
Web Services may seems to be a step backwards, yet document-oriented
services and REST are gaining popularity. The last 20 years have produced a long litany of technologies based on these concepts, including
ODBC, CORBA, DCE, DCOM, RMI, DAO, OLEDB, SQLJ, JDBC,
EJB, JDO, Hibernate, XML-RPC, WSDL, Axis and LINQ. Even with
these technologies, complex design patterns for service facades and/or
bulk data transfers must be followed to optimize communication between client and server or client and database, leading to programs that
are difficult to modify and maintain.

While significant progress has been made, there is no widely accepted
solution or even agreement about what the solution should look like. In
this talk I present a new unified approach to invocation of distributed
services and data access. The solution involves a novel control flow construct that partitions a program block into remote and local compu-
tations, while efficiently managing the communication between them.
The solution does not require proxies, an embedded query language, or
constructions/decoding of service requests. The end result is a natural
unified interface to distributed services and data, which can be added to
any programming language.

A. Dearle and R.V. Zicari (Eds.): ICOODB 2010, LNCS 6348, p. 8, 2010.

Keynote Panel "New and Old Data Stores"

Panelists:

- Ulf Michael (Monty) Widenius, main author of the original version of the open-source MySQL database.
- Michael Keith, architect at Oracle.
- Patrick Linskey, Apache OpenJPA project.
- Robert Greene, Chief Strategist Versant.
- Leon Guzenda, Chief Technology Officer Objectivity.
- Peter Neubauer, COO NeoTechnology.

Abstract. The world of data management is changing. The linkage to service platforms, operation within scalable (cloud) platforms, object-relational bindings, NoSQL databases, and new approaches to concurrency control are all becoming hot topics both in academia and industry.

The name NoSQL databases attempts to label the emergence of such growing number of non-relational, distributed data stores that often did not attempt to provide ACID properties. ACID properties are the key attributes of classic relational database systems. Such "new data stores" differ from classic relational databases, they may not require fixed table schemas, and usually avoid join operations and typically scale horizontally.

The panel discusses the pros and cons of new data stores with respect to classical relational databases.

A. Dearle and R.V. Zicari (Eds.): ICOODB 2010, LNCS 6348, p. 9, 2010.

Revisiting Schema Evolution in Object Databases in Support of Agile Development

Tilmann Zäschke and Moira C. Norrie

Institute for Information Systems, ETH Zurich
CH-8092 Zurich, Switzerland
{zaeschke,norrie}@inf.ethz.ch

Abstract. Based on a real-world case study in agile development, we examine issues of schema evolution in state-of-the-art object databases. In particular, we show how traditional problems and solutions discussed in the research literature do not match the requirements of modern agile development practices. To highlight these discrepancies, we present the approach to agile schema evolution taken in the case study and then focus on the aspects of backward/forward compatibility and object structures. In each case, we discuss the impact on managing software evolution and present approaches to dealing with these in practice.

1 Introduction

The introduction of agile software development methods impacts on schema evolution in at least two aspects. First, the agile preference for continuous evolution and refactoring of the data model results in more frequent schema evolution. Second, the shortened release cycles mean that not only the in-house data model has to be updated more often, but also the customers have to be more frequently provided with means to make their existing data usable with new software releases. Accordingly, and depending on the complexity of the data model and the environment, managing and implementing schema evolution can require significant effort in a software project.

The integration of traditional relational databases into object-oriented systems can restrict agility because of the need to maintain the object to relational mapping. The case has therefore been made that object databases are better suited to agile development methods [1]. Yet the support for schema evolution offered by the various object database products tends to be limited. As a result, application developers often have to produce significant custom code to manage evolution. At the same time, most of the solutions proposed in the research community precede modern software development practices and are based on invalid assumptions or address challenges that no longer represent the key issues.

To analyse the requirements and solutions for managing schema evolution, we examined the real-world case study of a software project that adopted agile development practices and uses an object database to store persistent data. The system was developed by the European Space Agency (ESA) to manage

A. Dearle and R.V. Zicari (Eds.): ICOODB 2010, LNCS 6348, pp. 10–24, 2010.

the scientific operation of their Herschel Space Observatory. It was found that little of the available research was applicable to the project, in part due to preconditions assumed (if not explicitly stated) by many researchers. While these preconditions are certainly valid for some projects, we believe that the features and requirements of the Herschel system are characteristic of many modern applications. Further, for some of the challenges faced by the developers of the Herschel system, we were unable to find any scientific references at all. The main contribution of this paper is to point out and give possible explanations for the differences between research and practice, thereby showing the demand and opportunities for further research.

We note that while our discussion is set in the Java/Versant world due to the case study considered, we believe that most of the lessons learned can be easily applied to other languages and object databases. Any parts of the discussion that are specific to the Java/Versant setting will be clearly indicated.

We begin in Sect. 2 with a discussion of the state of the art in terms of research and also support for schema evolution in object database products. An overview of the Herschel project is presented in Sect. 3 followed by a discussion of agile practices and their impact on requirements for managing schema evolution in Sect. 4. Sect. 5 discusses the issue of supporting backward and forward compatibility under schema evolution, while Sect. 6 examines the object structures supported in object database products and their effect on schema evolution. Sect. 7 provides a summary and discussion of our findings and concluding remarks are given in Sect. 8.

2   Background

In object databases, the schema evolution process consists of two main parts: evolution of the class schema followed by evolution of the object instances to reflect these changes. We will refer to these as *class evolution* and *data evolution*, respectively. Data evolution may require new values to be initialised or existing values to be restructured or transformed in some way and products typically offer an API which allows developers to write custom evolution code.

Schema evolution has been studied extensively over the years with many solutions proposed for both relational and object databases [14,13]. For object databases, two main categories of solutions have been proposed in research: a) methods that address the evolution of individual classes and b) methods to address the evolution of the entire database class schema.

In the first category, changes to a class include changes to attribute definitions, the creation or deletion of a class, and the creation or deletion of sublasses or superclasses. Such changes are no longer an issue in that most commercial object databases provide support for handling them, although the basic approach varies in terms of whether it is automated or under the control of the developer. For example, Objectivity[1] handles some changes automatically, while db4o[2] provides

---

[1] http://www.objectivity.com
[2] http://www.db4o.com

only an API that the developer can use to manage evolution. Versant[3] opts for semi-automatic support in that basic class changes can be handled automatically, whereas complex modification of the class hierarchy require the use of an API they provide.

The second category of solutions focus more on the evolution process and how to manage the evolution of the entire database class schema. For instance, in OTGen [8], the output of a schema change is a copy of the whole database for a new schema. In [5], Clamen proposes a functional mechanism to support schema evolution for centralised and distributed object databases that maintains compatibility for old applications. Other researchers have addressed the issue of backward compatibility of schema changes, e.g. CLOSQL [9]. However, as we will show later, this adds considerable complexity, especially in the case of agile development where evolution is frequent. Another major distinction between approaches is whether they version classes under evolution, see for example [15,3].

Although cycles in object graphs were used in the Herschel project, many research works assume directed acyclic graph (DAG) structures for both the class hierarchy and object graphs. Only a few publications discuss data structures including graphs with cycles. For example, [2] describes object structures with cycles but does not discuss the implementation of a schema evolution solution.

Similarly, the actual object structures supported in object database products cause some issues which have not been addressed by the research community. For example, several products introduce some concept of embedded or dependent objects to meet different requirements with respect to persistence. Specifically, many vendors introduce two categories of objects, first class objects (FCO) and second class objects (SCO). One of the key properties of SCOs that impacts on schema evolution is the fact that they do not have a class schema.

In summary, research has tended to focus mainly on primitives, rules, invariants and semantics of schema changes and less on managing the data evolution process. Further, much of this research was carried out before the adoption of agile development methods where evolution is much more frequent and one cannot assume that it is a case of evolving one complete and correct system into another. In the remainder of the paper, we will show how data evolution often proves to be the most complicated part, especially in very large systems under agile development.

3 Herschel System

The Herschel Common Science System (HCSS) is a project of the European Space Agency to support the scientific operation of the Herschel Space Observatory[4], an Infrared Observatory Satellite that was launched in May 2009. On the pre-observation side, the HCSS software provides the whole chain of submitting, evaluating and scheduling proposals for observation, instrument programming, editing of calibration tables and finally the generation of control commands for

---

[3] http://www.versant.com

[4] http://www.esa.int/science/herschel

the satellite. On the post-observation side, it is used for storing, extracting, calibrating, post-processing and distributing observational data. Excluding the post-processing, virtually all data is stored in object databases (Figure 1).

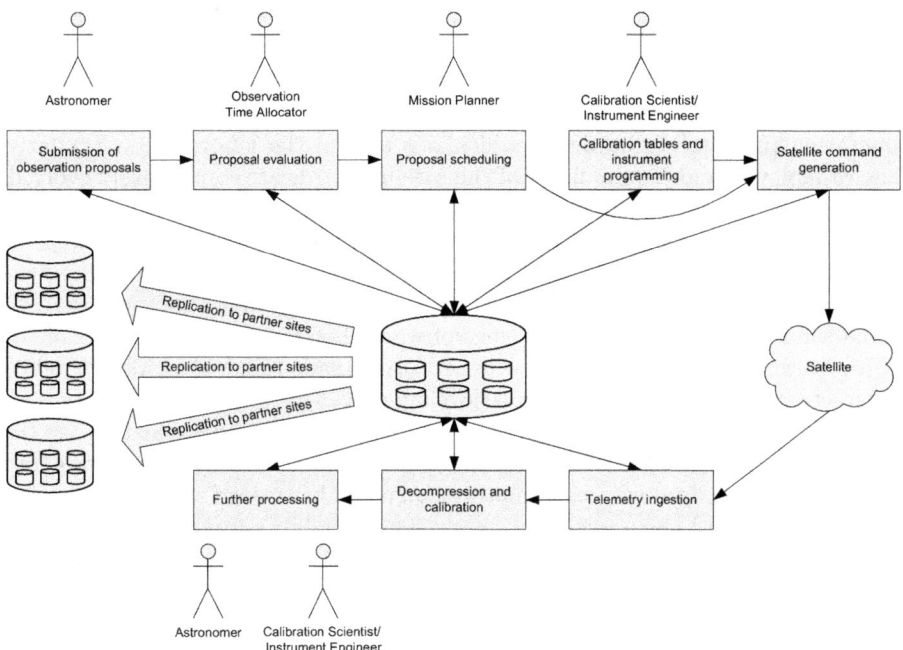

Fig. 1. Data flow between the subsystems of the HCSS software. All data (except for the last process) is stored in a central database system with multiple nodes. The database system is replicated to associated research institutions.

The operational time of the satellite is limited to 3-4 years, during which the expected data is in the order of $2*10^9$ objects which will amount to 10-15TB[5]. The data is partially replicated to other sites, so in total there are expected to be over 100 interconnected database nodes in the system. The software is implemented in Java 6, and the database used is the Versant Object Database accessed via Versant's Java interface (Versant JVI).

The project followed the agile manifest[6] quite closely, emphasizing continuous user involvement. In particular, the following agile practices used in the project were relevant to schema evolution: Continuous integration with builds being used by interested users, frequent user releases (6-8 weeks), partly test-driven development, continuous evolution and re-evaluation of design and requirements

---

[5] This depends on the usage of the different observation modes of the satellite, each of which produces data at a different rate.

[6] http://agilemanifesto.org

with close interaction between users and developers. A full description of all agile aspects of the project is outside the scope of this paper.

Agile development was the method of choice for this project for several reasons. Scientific satellites do not follow an off-the-shelf design but feature unique capabilities, requirements and operational concepts. Furthermore, HCSS is the first ever project in ESA using such a complete end-to-end software suite to operate a satellite. Therefore, there was little experience with similar software.

Starting with a general architecture and an initial set of user requirements, the software architecture evolved continuously as users gained experience with operating the satellite hardware which was still in the laboratories. Similarly, the capabilities and requirements of the satellite hardware continuously evolved, leading to more adjustments to the initial design and requirements of the software. The continuous use of early software releases by users (scientists) with their instrument hardware ensured that problems and any misunderstandings common to multi-party development[7] had been cleared and that users were familiar and satisfied with the resulting software. The fact that most parts of the software would only be used by few expert users made the project ideal for close interaction between these users and developers.

In summary, one major achievement of using agile concepts was that, at the time of launch, not only were the users highly experienced with the software, but also the software itself was really mature and could be trusted with the operation of a multi billion Euro space observatory. This reliability was especially critical because the lifetime of the satellite is strictly limited to 3-4 years[8], meaning that every lost day would have cost over one million Euro.

According to plan, development is now frozen for most parts of the system since launch, except for critical bug fixes. The only part still under development is the data processing module. Similar to the other parts of the software, which continuously evolved with the hardware and users gaining experience in hardware operation, the data processing module now evolves as users gain experience in processing the satellite imagery. Data processing was already under heavy development before launch, as it was also required for satellite hardware development and because it had to be sufficient to process the first incoming data after launch. However, since launch, the software has been developed further due, not only experience gained with real satellite imagery, but feedback from a much larger user community of astronomers who previously had little involvement in the project.

The use of agile development practices was not the only property of the project that was relevant to schema evolution. The size and complexity of the data structure with about 250 persistent classes, including cycles and redundancies, some of which have around $2*10^9$ instances, played a major role. Further, some large database systems consisted of up to 30-40 database nodes which were also replicated to other sites (shared administration). Other features of note are that

---

[7] USA and many ESA member states contributed software or hardware via national institutes.

[8] Operation is limited by the cooling liquid which is required for the infrared detectors.

the DBAs were mostly non-professional and regular down-times of databases and the system were acceptable.

In the following sections, we will first examine the impact of agility on schema evolution strategies and then discuss the issues of backward/forward compatibility and support for complex object structures in detail.

4   Agile Schema Evolution

In the HCSS project, the use of agile development methodologies had considerable impact on schema evolution. For schema versioning, the project chose a concept where only one version of a class schema was accessible in any database. To ensure schema compatibility, each database contained a *database schema ID* which defined the schema versions in the database. The database schema ID allowed connection only from a limited range of software versions with a correlating *software schema ID*. The compatible software typically comprised only one user release and correlating nightly builds. The mechanism with the single ID can also be used to prevent accidental connection by indicating an ongoing or failed schema update by setting the ID to a defined invalid schema version.

The schema evolution procedure was split into two parts, each performed by a dedicated application. The first application required database downtime to perform evolution of the class schema and critical data. The application also incremented the database schema ID. Where possible, the evolution of non-critical data was performed later by a second application which did not require downtime as it performed data evolution on a live database. Consequently, this requires clients to tolerate data that has not yet evolved.

The two applications were maintained by one dedicated developer who collected schema change requests and draft code for data evolution from other developers. According to a schedule, the developer would then implement and release the schema changes and the two schema evolution applications. The task of implementing schema evolution was simplified by the fact that all persistent classes were under the custodianship of the database developer.

In order to evolve a database, the DBAs had to execute the evolution applications on their databases. The challenges of implementing schema evolution applications included usability for clients with little database expertise, evolution of large databases, evolution of a database system comprising multiple dependent databases and the evolution of data. In particular, implementing the data evolution could be difficult. First, it had to account for sometimes very complex calculations of initialisation values for new fields based on data from many other objects. Second, it also had to deal with variably inconsistent databases caused by occasional bugs in the frequent user releases, occasionally used less stable nightly builds and even custom applications.

The effects, both negative and positive, of using agile development practices included:

Constantly evolving design. Constantly evolving design results in continually incoming requests for schema evolution. The HCSS project decided not to implement forward or backward compatibility between schema versions for reasons discussed in the next section, instead using the concept of schema IDs introduced earlier to enforce compatibility between a database and any application accessing it. Based on the schema change requests collected together with draft code for the data evolution, the developer responsible for schema evolution implemented the schema evolution tool and scheduled a release date.

Fast release cycle. To avoid holding back new features and fixes, schema evolution had to be performed at least once per user release. To accommodate the fast release cycle of only 6-8 weeks (Figure 2) and give developers time of uninterrupted development between releases, the schedule usually placed schema evolution in the week before the following code freeze. This left only one week for beta-testing the changes.

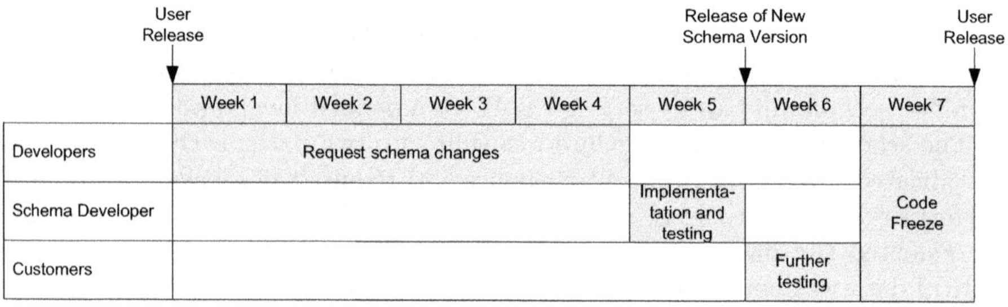

	Week 1	Week 2	Week 3	Week 4	Week 5	Week 6	Week 7
	↑ User Release				↑ Release of New Schema Version		↑ User Release
Developers	Request schema changes						Code Freeze
Schema Developer					Implementation and testing		
Customers						Further testing	

Fig. 2. Software release cycle with respect to the release of new schema versions.

Close interaction with users. During the one week of beta-testing, the close interaction with the users allowed them to run the new schema evolution code on copies of their most critical databases. The close interaction also made it possible to fix any issues that arose which would have not been possible otherwise.

Evolving incomplete or incorrect data. The beta-testing by users was essential because users often used nightly builds, patched builds and even custom software, which resulted in sometimes rare and unpredictable inconsistencies in their databases. Some inconsistencies would only occur in one or two of the more than 100 databases across the whole project, were virtually impossible to foresee and would not occur during previous internal tests by the schema evolution developer.

Flexibility. Another aspect of the close interaction with the users was the flexibility of the schedule, which was every time agreed with all developers and users, and had at times to be changed to avoid impact on other tasks.

Stacked evolution. The frequency of schema evolution and the fact that users sometimes skipped a release also meant that any schema evolution process had to support evolution of databases over several schema versions at a time (stacked evolution), without requiring the user to install any intermediate releases. At the same time, the software should not accumulate schema evolution code for every schema evolution, so it was decided to implement a central repository from which schema evolution applications would download evolution code for older databases, if required. Database servers that did not have internet connections were accommodated by making the repository portable to custom locations.

Maturity and training. By the time the project switched from the development phase to the critical operational phase, the procedure and user interface for schema evolution were matured and users were routinely applying the tools to their databases.

5   Forward and Backward Compatibility

Forward and backward compatibility for schema versions is a frequently discussed topic in research related to schema evolution. Forward compatibility allows old software to access a newer database, while backward compatibility allows new software to access an older database. Although compatibility and the resulting flexibility is an obvious advantage, the HCSS project chose not to support forward and backward compatibility for several reasons:

Old software accessing new data. Allowing old software to access a new database also implies that old known problems in that software continue to affect data in the database. In other words, locking out old software implicitly enforces a minimum patch level for all accessing applications.

New software accessing old data. Allowing a new software version to access old data means that new software always has to be prepared to encounter old inconsistent data that would have otherwise been fixed by schema evolution.

Forward compatibility. This requires additional implementation effort, because it is the inverse implementation of normal schema evolution.

Code cluttering. The code allowing backward compatibility is similar to the schema evolution code. However, having this code in the client has the disadvantage that there have to be as many versions of the code as there are schema versions. In the HCSS project, there were over 35 schema versions in 5 years. This can significantly affect performance if old objects have to be evolved over multiple versions.

Errors during data conversion. Failures in the forward or backward conversion code confront the end-user with unrecoverable application errors which can only be fixed by developers or DBAs.

Performance and quality of service. Loading objects from an incompatible schema version can impact performance in the case of complex evolution algorithms or algorithms that require loading of additional objects. The latter can even cause the data evolution code to hang or cause *other* applications to fail, because the additionally required objects may cause unexpected locking problems. These issues are aggravated by the possibility that some objects may need to be evolved over many schema versions. An additional problem is that these issues only occur for objects of particular schema versions, meaning that application performance can become unpredictable.

Forensics. Determining the cause of database inconsistencies gets considerably more complicated through the increased number of software versions accessing the database and through the different paths of data access, namely multi-level forward or backward conversion, or direct access.

Separation of concerns. From a design point of view, schema evolution code does not belong in the domain of user applications. The design in the HCSS project kept normal applications free from such code.

A direct consequence of separating evolution code from clients is the need for *standalone tools*, i.e. applications, for schema evolution. Besides avoiding the above problems, some advantages of standalone evolution tools are:

- By the time the tool is finished, no more failures can occur, because all objects are evolved.
- If the tool requires a downtime, it can take advantage of being the only application accessing the database. For example, disabling locking improves performance and simplifies the evolution code.

The HCSS project implemented two distinct schema evolution tools. The first requires database downtime and evolves the class schema and some of the data. The second tool runs later in parallel to other client applications and evolves remaining data.

To minimize the required downtime, the first tool uses a kind of *partially delayed class evolution*, known as lazy evolution in Versant terminology. This means that the tool can update a class that requires only a simple change such as adding a primitive attribute within less than a second, independent of the number of instances of this class. The instances are later evolved transparently when the objects are accessed by clients. This functionality allows only simple updates such as removing or adding an attribute initialised to '0'. Apart from the minimised downtime, there is virtually no impact on client applications, because they need no extra code, the initialisation to '0' is very fast, and it is unlikely to fail.

The second tool runs later in *parallel to other applications* and evolves non-critical data. Non-critical data refers to data where clients are either prepared to encounter non-evolved data, or where the client itself is not critical and can afford an extended downtime until data evolution is finished. In the HCSS project, this delayed data evolution was only used for data with large cardinality that would have otherwise increased the required downtime by several hours.

We conclude that including schema evolution code in client applications needs careful evaluation. Especially projects that can afford database downtime are likely to fare better with simpler solutions. We can see advantages of forward and backward compatibility for short-term use to support smooth transition between schema changes when the additional effort is justified by the possibility of avoiding database downtime altogether. However, our experience with the HCSS project showed that its simple approach can be superior given the right circumstances. In the HCSS project, even for larger databases, using the above concept with two distinct tools resulted in downtimes rarely exceeding a few minutes. So far, we could not find any research paper that discusses all of the above issues or gives recommendations for cases in which forward or backward compatibility should be implemented.

6   Object Structures

While most research tends to assume a relatively simple and pure object model, in practice, many object databases have constructs with special semantics, such as second class objects (SCOs) or non-DAG structures, that require special handling in schema evolution. We will first look SCOs and their counterpart first class objects (FCOs). The JDO specification [7] defines them as follows:

> "A First Class Object (FCO) is an instance of a persistence-capable class that has a JDO Identity, can be stored in a datastore, and can be independently deleted and queried. A Second Class Object (SCO) has no JDO Identity of its own and is stored in the datastore only as part of a First Class Object."

Versant uses an equivalent definition [16] in their (non-JDO) Java API, where *JDO Identity* is substituted by *Object Identity*. Objectivity supports a similar concept to SCOs for both Java [10] and C++ [11] called *embedded objects*. A common feature of SCOs is that they do not have an identity in the database, and their existence depends on an FCO or another SCO that references them. Note that primitive types, arrays and java.lang.String (in Java) are generally supported and not considered as SCOs.

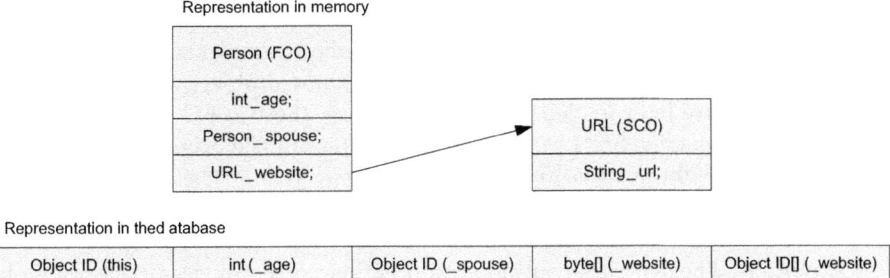

Fig. 3. FCO (Person) references SCO (URL)

Figure 3 shows an example of an FCO of class Person and an SCO of class URL. Person has a schema defined in the database, but URL has not. Looking at Versant, when a URL is stored in a database, it becomes part of the Person and is serialized into an array of bytes and an array of object IDs for storing possible references from the SCO to other FCOs. The relationship changes from aggregation in memory, where the URL is an independent object, to composition in the database which means that deleting the Person from the database will also delete the URL.

The properties of SCOs can be summarised as follows:

No schema. They do not have a schema defined in the database.

References to FCOs. When serialised, references from SCOs to FCOs are replaced by their respective object IDs. Referenced SCOs are serialised in-line.

Not managed. SCOs typically do not contain any database related code (inserted manually or automatically via byte-code enhancement (Java) or macros (C/C++)). This allows the storage of third party objects as SCOs in a database. One resulting issue is that they are stateless (see below).

Dependent objects. While stored in a database, they behave much like *dependent objects* [2] in that they can have only one owner, and their existence depends on that owner. Like dependent objects, they can reference other dependent or independent objects. Their owner can be an independent or dependent object. These restrictions do not apply to SCOs in memory.

Although common in practice, to the best of our knowledge, no research paper has yet fully discussed SCOs. In particular, we have not seen any mention of the implications of the above properties. Some implications and issues are:

Stateless / Not managed. SCOs do not have state flags and access is not monitored. For example, an SCO cannot be marked dirty, and modifying an SCO from unmanaged code will not flag its FCO as dirty either. Instead, an FCO has to manually be flagged as dirty whenever one of its SCOs is modified.

Queries and indexing. We do not know of any object database that supports indexes or queries on SCOs.

SCO duplication. SCOs are vulnerable to object duplication. Duplication occurs when an SCO is referenced by two FCOs. When the FCOs are stored, the FCO will be serialised twice in the database, resulting in two separate SCOs as shown in Fig. 4. This results in unwanted object duplication since when they are later loaded from the database, the two FCOs will no longer reference the same SCO but two distinct copies of the original SCO.

Uncontrolled schema evolution. One problem with SCOs that was encountered in the HCSS project occurred when a developer changed one of the serialised classes, but was not aware of the implications regarding schema evolution. The software continued to function correctly, because storing objects still worked and because the affected type of objects was rarely read. The problem was only discovered a few weeks later when someone tried to look at objects that were serialised before the class was changed and found that de-serialisation of these old objects failed. Versant provides no API dedicated to schema evolution of SCOs, but writing a custom de-serialisation class solved the problem.

A similar problem can occur when the SCOs stem from a third party library. Libraries usually only keep their API stable, not their internals. Therefore, any library update may contain changes in the fields of its classes. Furthermore, clients may have different versions of a library installed, with compatible APIs but possibly incompatible internal fields.

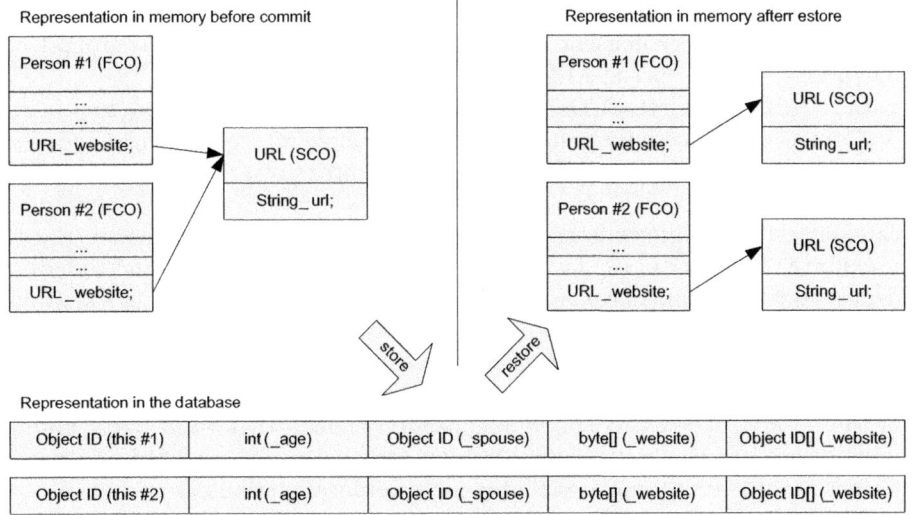

Fig. 4. SCO duplication during storage

False schema evolution. In Versant, only objects of classes that implement the `Serializable` interface can become SCOs. The catch here is that even changes that do not directly affect the schema, such as changing a field to `final`, can cause serialisation exceptions, because the `serialVersionUID` changes. This problem can be avoided by explicitly specifying this ID when the class is created, with the disadvantage that real schema changes may not be detected if the ID is not correctly updated.

Varying support. Support of SCOs varies between vendors such as Objectivity [10] (containers, `Date` and `Time`), Versant [16] (containers and classes that implement `java.io.Serializable` or `java.io.Externalizable`) and ObjectStore [12] (SCOs are completely prohibited).

Despite these problems, SCOs can be quite useful. There seem to be two main reasons for using SCOs.

The first reason is persistence of third party classes. Storing third party classes as SCOs directly in a database avoids having to map third party data to local classes. Often third party classes cannot be made persistent, mainly because the injection of management code (byte code enhancement in Java or macros in C/C++) is difficult or even impossible with third party libraries. Enhancing third party classes can also be problematic if the application requires these classes to have a certain structure, e.g. when serializing them into files or for network transport and communication with other applications. Using third party SCOs can also cause problems like *false schema evolution* described above.

The second reason for using SCOs concerns performance and storage efficiency. SCOs have no storage overhead and improve performance because they do not need unique object identifiers nor information about their location in a database. In the case of small SCOs, performance is also improved because SCOs provide a very efficient way of implementing clustering and group-load operations without special API calls.

In the HCSS project, SCOs were only used for the second reason, improving performance of numerous small objects. One example was the logging of function calls, with the parameters being stored as an array of type-value pairs, each pair being an SCO. We conclude that SCOs are useful and safe when used carefully. Our recommendations regarding SCOs in first party classes are to use serialVersionUID (in Java) to avoid *false schema evolution* and to have a dedicated developer manage SCO classes along with FCOs, possibly in a separate package in Java, to avoid accidental modification by other developers. For third party SCOs, we recommend ensuring that third party libraries can be strongly linked to the product so that users cannot accidentally load a different version of the library. It can also not be emphasized enough that third party library updates need to be checked very carefully and that a plan should be prepared for performing library updates that cause SCO incompatibility.

Two other common features of object structures encountered in practice have received little or no attention by the research community. The first of these concerns the general assumption in most fundamental research on schema evolution (for example [17]) that the *class hierarchy* is a DAG structure with all of its complexities. While single inheritance is technically covered by this research, no research investigates how much schema evolution is simplified by languages like Java that do not allow multiple inheritance.

As mentioned previously, many researchers also assume that the *object graph* is a DAG structure. While DAG-like data structures are clean and may be used in some projects, cycles and other redundancies are often introduced for pragmatic reasons. The HCSS project frequently used cycles to provide short cuts for navigation for applications with differing access patterns and to simplify queries with joins. Similarly, the HCSS project sometimes stored the same data in fields of different classes to avoid navigation altogether, thereby allowing for better indexing or avoiding joins in queries.

7   Discussion

As we have shown, the use of agile development methods has a strong impact on schema evolution. The issues resulted from the increased frequency and amount of schema evolution requests, from reduced time for implementing and testing schema evolution code and from the widespread use of nightly builds and custom applications, which lead to unpredictable inconsistencies in the databases and further complicated the testing of schema evolution code.

These issues were mitigated by other aspects of agile development and by additional techniques to manage schema evolution, for example the close interaction with end users which allowed for regular and efficient beta-testing of schema evolution code. Also, the use of the two distinct schema evolution tools proved to be very helpful, especially because their simple handling posed little problems to users. Finally, the policy of accumulating changes and releasing them at a regular, yet flexible, interval reduced the impact on other tasks.

Regarding the often proposed concept of forward and backward compatibility, we found that it can result in considerable problems when combined with agile techniques. The HCSS project did not implement forward and backward schema compatibility, instead using a simple yet powerful one-version concept, which was shown to be advantageous in all respects except the need for downtime. The advantages can be summarised as reduced development effort, better and more predictable application performance and the fact that only DBAs will be faced with schema evolution problems. Here we see an opportunity for further research that examines the impact of agile development methods on projects that use object databases. Moving away from backward and forward compatibility, this research could investigate alternative concepts and compare their fitness for different project settings.

Due to the use of SCOs and non-DAG structures, many schema evolution solutions proposed in research were not applicable. While their use may appear unorthodox, we believe that their benefits cannot be ignored in real-life projects like the HCSS. SCOs need careful handling, but can improve performance significantly. Non-DAG structures could be exploited to greatly simplify access code and improve performance and did not cause notable complications during schema evolution. We believe that the use of both SCOs and non-DAG structures deserve and could benefit from further research.

Considering the implementation of schema evolution code, another opportunity for further research would be the implementation of tools and a standardised API[9] for schema evolution, possibly embedded in an IDE. Contrary to frameworks proposed in earlier research [4,9,17,6], it would focus on the challenges of agile development and provide a standardised solution for multiple major vendors. We are currently planning to design such a framework based on our experiences from the HCSS project. We also plan to investigate improved support for SCOs and similar concepts, multi-node schema evolution and, depending on requirements, a system for deployment of stand-alone schema evolution applications as used in HCSS.

Finally, we note that in HCSS data evolution tended to be considerably more complex than class evolution, but is an issue that has received much less attention in research.

8   Conclusion

The HCSS project shows that agile development methodologies have a strong impact on schema evolution. Some agile aspects clearly increase schema evolution effort, while other aspects and good management practices help mitigating the effect. Considering technical decisions like forward and backward compatibility or details of object structures, we found that many of the questions faced during

---

[9] The latest JDO draft 2.3 proposes such an API.

the HCSS project are at most only briefly discussed in existing research literature and provide opportunities for further research.

Acknowledgements

We would like to thank Tilmann Zäschke's former employer VEGA[10], his former colleagues at ESA[11] who worked on the HCSS project and of course ESA itself, who own the HCSS project.

References

1. Ambler, S.W.: Agile Techniques for Object Databases (September 2005), http://www.db4o.com/about/productinformation/whitepapers/
2. Banerjee, J., Chou, H.T., Garza, J.F., Kim, W., Woelk, D., Ballou, N., Kim, H.J.: Data Model Issues for Object-Oriented Applications. ACM Transactions on Information Systems 5(1), 26 (1987)
3. Bjornerstedt, A., Britts, S.: AVANCE: an Object Management System. ACM SIGPLAN Notices 23(11) (1988)
4. Breche, P., Ferrandina, F., Kuklok, M.: Simulation of Schema Change Using Views. In: Proc. 6th Int. Conf. on Databases and Expert Systems Applications (1995)
5. Clamen, S.M.: Schema Evolution and Integration. Distributed and Parallel Databases 2(1) (1994)
6. Claypool, K.T., Jin, J., Rundensteiner, E.A.: Serf: Schema evolution through an extensible, re-usable and flexible framework. In: Proc. of the 7th Int. Conf. on Information and knowledge management. ACM, New York (1998)
7. Java Data Objects Expert Group. Java Data Objects 2.2. Technical Report JSR 243, SUN Microsystems Inc. (2008)
8. Lerner, B.S., Habermann, A.N.: Beyond Schema Evolution to Database Reorganization. In: Proc. ECOOP (1990)
9. Monk, S., Sommerville, T.: Schema Evolution in OODBs Using Class Versioning. ACM SIGMOD Record (1993)
10. Objectivity, Inc. Objectivity for Java Programmerís Guide Release 9.4 (2007)
11. Objectivity, Inc. Objectivity/C++ Programmerís Guide Release 9.4 (2007)
12. Progress Software Corporation. PSE Pro for Java User Guide Release 7.1 (2008)
13. Ram, S., Shankaranarayanan, G.: Research Issues in Database Schema Evolution: The Road Not Taken. Technical report, Boston University School of Management, Department of Information Systems (2003)
14. Roddick, J.F.: A Survey of Schema Versioning Issues For Database Systems. Information and Software Technology 37(7) (1995)
15. Skarra, A.H., Zdonik, S.B.: The Management of Changing Types in an Object-Oriented Database. In: Proc. OOPSLA 1986 (1986)
16. Versant Corporation. Java Versant Interface Usage Manual Release 7.0.1.4 (2008)
17. Zicari, R.: Primitives for Schema Updates in an Object-Oriented Database System: A Proposal. Computer Standards & Interfaces 13(1-3), 271–284 (1991)

---

[10] http://www.vegaspace.eu

[11] http://www.esa.int

The Case for Object Databases in Cloud Data Management

Michael Grossniklaus

Dipartimento di Elettronica e Informazione, Politecnico di Milano
P.za Leonardo Da Vinci, 32
I-20133 Milano, Italy
grossniklaus@elet.polimi.it

Abstract. With the emergence of cloud computing, new data management requirements have surfaced. Currently, these challenges are studied exclusively in the setting of relational databases. We believe that
there exist strong indicators that the full potential of cloud computing
data management can only be leveraged by exploiting object database
technologies. Object databases are a popular choice for analytical data
management applications which are predicted to profit most from cloud
computing. Furthermore, objects and relationships might be useful units
to model and implement data partitions, while, at the same time, help-
ing to reduce join processing. Finally, the service-oriented view taken by
cloud computing is in its nature a close match to object models. In this
position paper, we examine the challenges of cloud computing data management and show opportunities for object database technologies based
on these requirements.

1 Introduction

Database management systems are used in a wide variety of applications, ranging from mobile or embedded scenarios to large-scale solutions to support data-
intensive and global applications. To address the requirements of different applications, different database technologies have emerged and the consensus today
is that "no size fits it all" [1]. Therefore, one of the challenges has become to
match technologies to requirements. The vision of cloud computing is to solve this
problem by making computing a commodity that adapts to initial application
requirements, but can also evolve and gracefully scale when these requirements
change.

While people from different fields have slightly different definitions of the term
Cloud Computing[1], the common denominator of most of these definitions is to
look at processing power, storage and software as commodities that are readily
available from large infrastructures and, thus, no longer have to be provided by
desktop computers or local servers. As a consequence, cloud computing unifies elements of distributed, grid, utility and autonomic computing to provide software,

---

[1] Multiple Experts Try Defining "Cloud Computing":
    http://tech.slashdot.org/article.pl?sid=08/07/17/2117221

A. Dearle and R.V. Zicari (Eds.): ICOODB 2010, LNCS 6348, pp. 25–39, 2010.

platforms and infrastructure as a service. At the lowest level, Infrastructure-as-a-Service (IaaS) offers resources such as processing power or storage as a service. Examples include Amazon's Elastic Compute Cloud (EC2)[2], Sun Cloud[3] and GoGrid[4]. One level above, Platform-as-a-Service (PaaS) provides development tools to build applications based on the service provider's API. Notable solutions on this level are Microsoft's Windows Azure Platform[5] and the Google App Engine[6]. Finally, on the top-most level, Software-as-a-Service (SaaS) describes the model of deploying applications to clients on demand.

The impact of cloud computing on data management research, and on query processing in particular, has recently been studied by Abadi [2] and Gounaris [3], respectively. Abadi argues that characteristics such as scalability through parallelization, storing data on untrusted hosts, and wide geographic distribution or replication, render cloud computing unsuitable for transactional data management. These applications are typically quite write-intensive and require strict ACID guarantees. Both of these characteristics do not match well with the properties of a cloud computing environment. However, analytical data management applications that mostly query large data stores for decision support or problem solving will profit from these properties. Further, this type of application is also becoming increasingly important both in science and industry [4]. Both authors agree that the requirements of data management in cloud computing can be partially addressed by integrating existing results from database research into systems that combine features from parallel, distributed and stream database management systems. However, they also point out that, in order to deliver on the cloud computing vision, hybrid solutions that integrate other execution paradigms have to be considered to better support complex analytic and extract-transform-load tasks [5].

The direction into which cloud computing data management is evolving, makes it an ideal setting to investigate the use of *object databases* as they have become a popular choice for data-intensive analytical processing tasks. For instance, the Herschel project[7] of the European Space Agency (ESA) uses the Versant Object Database[8] to store, manage and process all data gathered by the telescope in outer space. Another example is Objectivity/DB[9] that is often selected for analytical scenarios such as the Space Situational Awareness Foundational Enterprise (SSAFE) that tracks space debris in real-time to avoid collisions in future space missions. Additionally, Objectivity has recently released a version of their product that can be deployed on cloud computing infrastructures.

---

[2] http://aws.amazon.com/ec2/
[3] http://www.sun.com/solutions/cloudcomputing/
[4] http://www.gogrid.com
[5] http://www.microsoft.com/windowsazure/
[6] http://appengine.google.com/
[7] http://www.esa.int/herschel
[8] http://www.versant.com
[9] http://www.objectivity.com

In this position paper, we will examine the exact requirements of cloud computing data management and, based on these challenges, demonstrate the corresponding opportunities for and benefits of object databases. We begin in Sect. 2 by introducing the three main challenges of cloud computing data management. The current state of the art in cloud computing data management is highlighted in Sect. 3. As this body of research is very vast, we have decided to point out the most influential approaches, rather than to give a comprehensive survey. In Sect. 4, we revisit the three main challenges and show how object database technologies can be leveraged to address these requirements. We conclude in Sect. 5.

2   Challenges of Cloud Data Management

The challenges of cloud computing data management can be summarized as massively parallel and widely distributed data storage and processing, integration of novel processing paradigms as well as the provision of service-based interfaces. In the following, we will examine each of these challenges in more detail.

2.1   Parallel and Distributed Data Storage and Processing

One of the promises of cloud computing is to make computing power a commodity that can easily scale based on the current requirements of the application. This elasticity is typically achieved by allocating more resources in terms of servers, processor cores or memory and storage space to an application. As a consequence, leveraging these additional resources is more complex than migrating an application to a single more powerful machine. As pointed out in [2], the workload of an application has to be parallelizable in order to profit from cloud computing. In the context of data management, this means that applications which have been designed to run on a shared-nothing architecture [6] are good candidates for cloud computing.

Replication and distribution of data are further important characteristics of cloud computing data management [2]. Apart from delivering scalable computing power, reliability and quality of service are important goals of cloud computing. In terms of data management, this translates to providing highly available and durable cloud storage solutions. On the one hand, replication on a large geographic scale serves a dual purpose in this setting. First, the possibility to transparently replicate data ensures availability and durability. Second, a global network of cloud storages also allows the computation to be moved close to the client. Distribution, on the other hand, will arise as a natural consequence of the service-oriented architecture of cloud computing. In this service-based setting, it is foreseeable that a loosely coupled form of distribution will have to be supported, where many of the traditional assumptions about the data schema, distribution and statistics are no longer guaranteed to hold.

As we will discuss in the next section, both parallelism and distribution are topics that have been studied in detail. While these results are a good starting point, their applicability in widely distributed and heterogeneous settings is limited [3]. As a consequence, the challenge of parallelism and distribution in cloud computing data management lies in developing paradigms and technologies that are capable to address the requirement of massively parallel and widely distribution data processing and storage.

2.2   Integration of Novel Processing Paradigms

Data stream management systems [7] have introduced a processing paradigm that is different from the one of traditional database management systems. Instead of dynamically running queries over mostly static data, data stream management system register static queries over dynamic data. This property has made them a successful choice in the analysis of large volumes of rapidly changing data that arise, for example, in the real-time monitoring of complex systems. Most data stream management systems operate by evaluating the registered queries over so-called windows that extract a finite set of data out of an otherwise infinite data stream. Depending on the systems capabilities, these windows are advanced based on their size or at predefined intervals and the results can be recomputed or maintained incrementally. As applications in this domain are becoming more important, data processing in cloud computing needs to be applicable to both stored and streaming data [3].

By its very nature, cloud computing takes a very service-centric view on computing. In terms of data management, this signifies that database management systems will only be accessible through service-based interfaces. This development can already be observed on the Web today as many data sources are exposed using interfaces such as REST or SOAP, rather than a traditional database interface. The consequences of this evolution to data management and, in particular, data processing are manifold. In traditional database management systems, the performance of query processing depends on a number of factors. At query compilation-time, precise statistics about data distribution are required by the optimizer in order to determine the evaluation order of the query and chose the most appropriate physical implementations of the logical operators. At run-time, the query execution time is additionally governed by the storage layout, data clustering and access methods such as indexes. While traditional query processing is by no means trivial, the problem becomes even more complex in the setting of service-based data sources. As the data itself might not be under the control of the data management system, it is difficult to obtain complete and reliable data statistics. However, as query execution time is mainly dominated by the time it takes to invoke a service and the distribution of the data it returns [8], precise statistics are even more important. A successful cloud computing data management system will need to address these challenges in order to leverage data sources with service-based interfaces.

Finally, the paradigm of map-reduce [9] has recently been proposed to execute massively parallel data processing tasks. The name "map-reduce" stems from the fact that these systems decompose processing tasks into a map and a reduce step. Both functions are provided by the client of the system and defined based on a data set that is given as ⟨key, value⟩ pairs. The map function is defined as

$map(k_i, v_1) \rightarrow \texttt{list}(k_j, v_j)$, i.e. it takes an input data tuple of one domain and returns a list of output data tuples of another domain. In the first processing step, a map-reduce system applies the map function to all input in the data set. It then collects the resulting lists of output tuples and regroups them according to keys. Then the reduce function, defined as $reduce(k_2, \texttt{list}(v_2)) \rightarrow \texttt{list}(v_3)$, is applied to each group producing zero or more result values. Finally, these values are collected in one list and returned to the client. Typically, the map operation works as a kind of fork that splits the data set into smaller pieces, while the reduce operation is comparable to a join or aggregation that reassembles the final result. Therefore, the authors of [5] argue that map-reduce is more similar to extract-transform-load systems than to a database system. However, since neither map-reduce nor traditional parallel database systems can deliver the full promise of cloud computing data management, there is agreement that hybrid solutions that integrate both processing paradigms have to be built [2].

2.3 Provision of Service-Based Interfaces

In Sect. 2.2, we have reasoned that a cloud computing data management system has to be able to process tasks over service-based data sources. In this section, we make the complementary argument and motivate why cloud computing data management systems themselves have to offer a service-based interface.

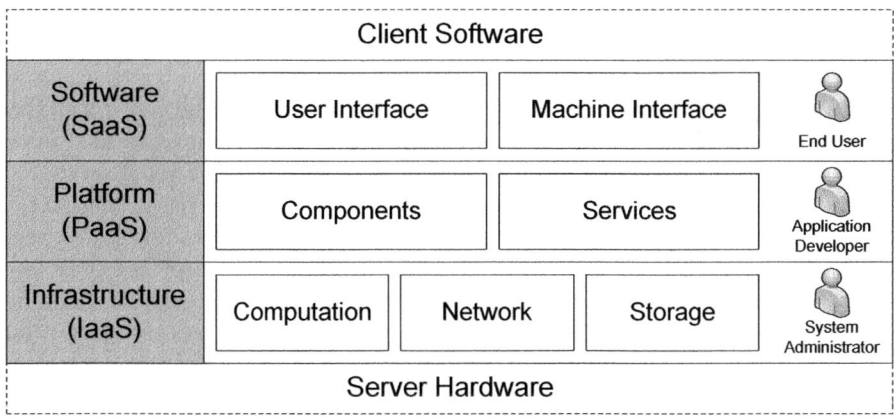

Fig. 1. Cloud computing stack, adapted from Wikipedia (`http://www.wikipedia.org`)

Figure 1 shows an overview of the cloud computing stack as it is generally agreed upon. The lowest layer, Infrastructure-as-a-Service (IaaS), provides computing resources such as processing power, network and data storage as services. The main goal of this layer is to deliver on the previously mentioned promise of providing elastic computing resources that scale gracefully in correspondence to the application's demand. Typically, this elasticity is achieved using platform virtualization. As a consequence, scalability signifies more resources, i.e. more processor cores, more network bandwidth and more storage space, rather than migrating an application to a more powerful machine behind the scenes. The IaaS layer is built on top of the server layer that consists of hardware products that support the delivery of cloud services through technologies as multi-core processors and hardware virtualization.

The aim of the Platform-as-a-Service (PaaS) layer is to provide an integrated computing platform that facilitates the development and deployment of applications. As the PaaS layer is built on the service-based interfaces of the IaaS layer, PaaS solutions typically offer software components and high-level services that can be used to implement applications. Examples of high-level services include relational database engines, payment services or costumer relationship management. Apart from development support, some PaaS solution also offer application hosting in the sense that an application that has been implemented using the offered components and services can be deployed to the vendor's cloud computing infrastructure.

Finally, the Software-as-a-Service (SaaS) provides end-user applications as a service. The advantages of delivering software as a service over the Internet are numerous. SaaS eliminates the need for customers to install the software on their local machines, while, at the same time, always working with the most up-to-date version of the application. For the software manufacturer, the advantage of SaaS lies in better control over the licenses that are in use and prevention of illegal duplication of the application. Cloud application services are built using the service-based interfaces of the PaaS layer. Furthermore, they typically expose service-based interfaces themselves in order to enable application interoperability.

With respect to data management, we note that there are challenges on all three layers of the cloud computing stack. Cloud computing data management starts on the IaaS layer with the provision of appropriate storage management services. In particular, the challenge on this layer is to provide a service-based interface for the functionality that corresponds to the physical layer of a traditional database management system. A declarative interface to the cloud computing data management system that facilitates data definition, manipulation and querying would typically be situated on the PaaS layer. In particular, data processing functionality such as traditional query evaluation or the aforementioned novel paradigms will be realized on this layer and offered using service-based interfaces to upper layers. Finally, user interfaces such as database browsing, designing or administration applications will be provided on the SaaS layer. Apart from these generic tools, also custom end-user applications will be deployed and hosted on this layer. As a consequence, a cloud computing data management system needs to address the challenge of providing service-based interfaces at various levels. Furthermore, system components located at higher layers need to be capable of leveraging and orchestrating the services on lower layers to implement their functionality.

3   State of the Art

Since cloud computing data management is a rather new discipline, the current state of the art is still quite limited. However, several works exist that identify challenges of cloud computing data management [2,3]. The requirement of storing and processing data in a parallel and distributed setting has been addressed in several existing works. On the one hand, research on *parallel databases* has led to a good understanding of parallelism, both in architectures [6] and query processing [10]. Research on *distributed databases* [11], on the other hand, has lead to results that can be leveraged to address distributed query processing [12].

Due to the focus on analytic data management, processing tasks in cloud computing are expected to be complex and long-running. Therefore, another requirement is fault tolerance in terms of self-optimizing and self-healing systems. This requirement relates cloud computing data management to the field of *autonomic computing* [13] and corresponding approaches to query processing (e.g. [14]). Another field of interest is *adaptive query processing* [15], where adaptive execution models (e.g. [16]) have been developed that demonstrate how entire query plans can be continuously adapted to available resources. The same goal is attained by adaptive query operators (e.g. [17]) that provide adaptation within individual nodes of an otherwise static query plan.

Data stream management systems [7] (e.g. [18]) are often mentioned for their capability to process rapidly changing data sets. However, it has to be noted that their data processing paradigm is fundamentally different from traditional query processing. As discussed in [3], significant advances have been made in this area in terms of query optimization, but they do not yet extend to the widely distributed and massively parallel setting of cloud computing.

With respect to integrating traditional data processing and service-oriented computing, approaches such as query processing over Web services [19,20] or search computing [21] have also to be considered relevant. In [19], a general-purpose Web Service Management System (WSMS) is described that can optimize and execute select-project-join queries spanning multiple Web services. The authors of [20] propose the "Serena" (service-enabled) algebra and a corresponding execution environment. Serena is based on the relational algebra, extended with service calls that can either be "get" or "set" calls and are classified according to whether they have or do not have side-effects. The authors also present rewrite rules for Serena that form the basis of rudimentary optimizations. Finally, search computing [21] extends query processing over services to search engines that return ranked results. The query processor developed in the project follows a traditional database approach in the sense that declarative queries are transformed into a logical plan. This plan is then optimized and translated into an executable physical plan. The optimizer choses the best query plan using a branch-and-bound algorithm that uses heuristics for determining the plan topology and load-balancing. Similarly to a traditional databases management system, the query execution environment supports different implementations of the logical operators such as join strategies that govern in which order the results from two search services are combined.

The field of *grid computing*, which can be considered a predecessor to cloud computing, also takes a service-centric approach on computing. Data management in grid computing has been studied intensely by several surveys (e.g. [22]) and research in this field has yielded results both in grid database systems (e.g. [23]) and grid query processing (e.g. [24]).

MapReduce [9] and related software are designed with built-in fault tolerance and capable of processing massively parallel and complex execution tasks at a large scale. Early approaches that point into the direction of integrating database functionality and map-reduce software have already been proposed, e.g. Yahoo's Pig Latin [25] or Microsoft's DryadLINQ [26], and SCOPE [27]. These approaches are only able to integrate the two paradigms at the language and not at the system level, since they layer map-reduce interfaces on top of traditional parallel databases. The approach of a hybrid architecture that supports multiple paradigms side-by-side is taken by HadoopDB [28] and Clustera [29], an integrated computation and data management system that is capable of executing database queries, workflows over Web services and map-reduce processes. In [30], a benchmark for large-scale data analysis systems is defined and implementation concepts for future hybrid systems are recommended.

In summary, numerous approaches exist that contribute to addressing these requirements. However, the core challenge of extending and integrating them into a comprehensive platform has not been fully addressed so far. Furthermore, most of these existing works were conducted in the context of the relational model. We believe that it is also imperative to study the possibility to exploit technologies from *object databases* [31] for cloud computing data management. This claim is motivated by the above-mentioned observation that analytical data management applications will benefit most from cloud computing and the fact that this class of applications is an ideal use case for object databases. Existing results from the domain of object databases that are relevant in this context include the works on object algebras (e.g. [32]), query processing (e.g. [33]), and query optimization (e.g. [34]). Further, approaches such as OMS Connect [35] have shown how features unique to object databases can be used to support multi-databases and modular peer-to-peer databases.

4 Opportunities for Object Databases

Object databases are a good match to both the type of data management applications that is anticipated to benefit most from cloud computing and the service-oriented view taken by cloud computing. As a consequence, we believe that cloud computing research needs to include these technologies. In order to make the case for object databases in cloud computing data management, we will now revisit the challenges outlined in Sect. 2 and show possible opportunities for object database technologies.

4.1   Parallel and Distributed Data Storage and Processing

As explained in Sect. 2, cloud computing takes parallel and distributed data storage and processing to a new level by requiring it to be massively parallel and widely distributed. In the following, we will first examine the case of data storage before looking at the case of data processing. Typically, parallel and distributed data storage is addressed through horizontal and vertical partitioning of the dataset.

In the setting of parallel databases that use the relational model, tables can be partition horizontally using selection predicates that segment a table into smaller ones. These table segments are then placed on different computing nodes and thus provide support for parallel processing of data manipulation operations. In contrast to the relational model, object data models provide more opportunities for application developers to define horizontal partitions. While the value-based approach is still possible, object databases can also leverage the existence of class extents or object collections for horizontal partitions. For example, Objectivity/DB and ObjectStore both support concepts that support the explicit clustering of objects into "containers", i.e. object collections that also govern the physical storage layout. Furthermore, Objectivity/DB is built around the concept of federated databases that could prove helpful in realizing such dataset partitions.

A vertical partition in a relational database management system segments a table in terms of columns, i.e. a (not necessarily strict) subset of the columns of a table are placed on different computing nodes. Object data models also provide ampler possibilities to realize vertical partitions. For example, the object-slicing technique [36] could be used to partition classes by leveraging the inheritance hierarchy and to distribute object data accordingly. Additionally, the existence of references and relationships between objects is a valuable asset to partition a dataset vertically as they can serve as natural points of decomposition [35]. As most existing object databases support binary relationships that are managed independent of the objects themselves, references can easily be traversed in both directions and thus bridging different partitions is straightforward.

In the past, several techniques (e.g. [37]) have been proposed to support horizontal, vertical and method-induced class partitioning in object databases. However, the requirements of cloud data management raise the question whether these partitioning schemes are still sufficient or whether more advanced techniques are required. For example, a recent article in InformationWeek[10] discusses the adoption of cloud computing in industry. The authors state that companies are attempting to split their data management needs between in-house and cloud computing platforms. This new form of partitioning allows transactional and analytical processes to be delegated to the appropriate computing infrastructure. We believe that also in this setting, objects and relationships are a useful unit to model, support and bridge such partitions.

---

[10] http://www.informationweek.com/news/showArticle.jhtml?
articleID=221901196

According to [11], the cost of data processing in parallel and distributed databases is generally a weighted combination of disk I/O, CPU and communication cost, where communication cost is typically considered as the most important factor. Therefore, query operators that require access to data from different partitions are typically associated with high costs. For example, in the case of vertical partitions, a processing node that executes a relational join of two tables or table segments has to access both operands. If the two operands are not stored on the same node, such a join operation can be very expensive. In the setting of object databases, it is again possible to profit from the existence of, potentially bidirectional, references. In contrast to the relational join operation that computes relationships between data at query execution time, references are managed statically by the database. Therefore, they do not have to be materialized, but can simply be navigated without accessing the referenced object. The benefits of leveraging references or pointers in parallel and distributed object databases have been demonstrated in the past. For example, the authors of [38] present several parallel pointer-based join algorithms for set-valued attributes, together with an evaluation of their performance. Similarly, ParSets [39] have also been shown to increase the performance of object graph traversals through parallelization. In the case of horizontal partitions, the costly operation is the one that performs a union over the data segments that contribute to the query result. However, in contrast to the traversal of relationships, this operation is typically less costly as a query optimizer can avoid to access remote data that does not contribute to the final result. Nevertheless, we point out that most object databases already feature collection data structures including the associated collection operations such as union, intersection and difference [40].

4.2   Integration of Novel Processing Paradigms

As object databases are situated at the intersection of object-oriented programming languages and database management systems [31], they are already tightly integrated with programming languages. As a consequence, the separation between the language and the system level is less pronounced. In fact, many object databases rely on programming rather than dedicated query languages to specify processing tasks. In the following, we will examine the characteristics of object databases that facilitate the integration of the processing paradigms introduced in Sect. 2, i.e. data streams, services and map-reduce.

The integration of data stream and traditional data management is difficult because the two processing paradigms are fundamentally different. In traditional data management, various dynamic queries run over a slowly changing database, whereas in data stream management queries are statically registered and process rapidly changing streams. Regardless of the fact that early data stream management systems have been proposed over a decade ago, the processing of streaming object data has not yet been investigated. Nevertheless, object databases are a suitable candidate for the integration of these two paradigms. On the one hand, the fact that classes of objects can define methods and the object database supports their execution provides a mechanism to "register" queries. On the other hand, the presence of events and listeners as, for example, in the Versant Object Database forms the basis for event-based processing which can be applied to realize data stream management.

To process data over service-based data sources, data management systems need to address the issues of long access times for data and uncertainty because of lacking data statistics. Furthermore, services have, in contrast to other data sources, interfaces that distinguish between ingoing and outgoing fields. As a consequence, selections and join predicates can be delegated to the services themselves whenever constants or outgoing fields of one service can be matched to ingoing fields of another. This property of service-based data processing gives rise to two types of joins, namely parallel and pipe joins [21]. In a parallel join, two services are invoked at the same time and the returned results are combined using a join predicate. This type of join corresponds to a relational join and the fact that services are invoked in parallel reduces execution time. If, however, the overlap in the output of the two services is small, it can also lead to costly and superfluous service invocations. In a pipe join, one service is invoked first and its output is used as input for the second service. While the sequential invocation of services might be less efficient, it allows the second service to be queried in a more directed way. This second type of join is similar to index-based joins or the notion of navigating object references. Therefore, object databases also provide a good basis for the integration of service-based data processing with traditional data management.

As mentioned before, map-reduce is a paradigm which provides a simple model that allows complex distributed processes to be specified. One of the advantages of map-reduce is that the base data (e.g. Web pages) can be cast into different implicit models such as bag of words, set of paragraphs, set of links, or list of links. The disadvantage of this approach is that there is no type checking during query processing since the model or type is constructed on the fly. Object databases could be used to support typing of queries by defining different object wrappers for the same base data instances. However, database and so-called extract-transform-load systems have very different architectures which makes their integration challenging. Early hybrid approaches can be classified into vertical architectures that build higher-level map-reduce interfaces on top of existing database systems and horizontal architectures where the two paradigms exist in parallel. However, as most object databases are already tightly coupled with object-oriented programming languages, they present a unique opportunity to investigate the integration of further processing paradigms. One possible approach to do so is to extend an object-oriented programming language with a domain-specific component that is handled by a compiler plug-in. In this way, DryadLINQ [26] has integrated map-reduce at the language level in the same way as LINQ has extended C# with query capabilities. Another interesting direction is to investigate object query languages that already provide operations similar to map and reduce as, for example, the algebra associated with the OM data model [41].

4.3 Provision of Service-Based Interfaces

In Sect. 2.3, we have motivated the challenge of providing service-based interfaces at all layers of the cloud computing stack as the complementary challenge to support data management over services. The limitation of relational database management systems in this context is twofold. On the one hand, services and relational data management are difficult to integrate as the two models do not align well. On the other hand, using relational systems and the software stacks that surround them in a service-oriented architecture is challenging due to the complexity of mapping service-based interfaces to the relational model. In a sense, this criticism goes back to the original impedance mismatch between object-oriented systems and relational databases [42], with the difference that it nowadays also applies to service-oriented architectures. With the the large-scale deployment of service-based data sources that is to be expected in the setting of cloud computing, the object-relational mapping overhead will grow to new dimensions, too.

As object data models and service-oriented interfaces are closely related, we are convinced that object databases have a lot to offer to cloud computing data management. The concept of orthogonal persistence, that is an essential feature of most recent object databases, is particularly relevant in this context. For example, the authors of [43] point out that the use of orthogonal persistence can already be observed in many modern systems. They speculate that the notion of orthogonal persistence could be extended in order to simplify the development of cloud applications. Instead of only abstracting from the the storage hierarchy, this extended orthogonal persistence would also abstract from replication and physical location, giving transparent access to distributed objects.

5   Conclusion

The promise of cloud computing to render computing a commodity is a promising direction that should also include data management capabilities. In this paper, we summarized the challenges that are associated with delivering cloud computing data management. We argued that data management solutions in the cloud need to be capable of storing data massively parallel and widely distributed. Further they need to integrate novel processing paradigms, such as data stream, service-based and map-reduce processing. Finally, cloud computing data management systems must themselves provide service-based interfaces in order to integrate horizontally and vertically in the cloud computing stack. While some of these challenges have previously been identified by other authors [2,3], this paper presents an integrated and extended view of the requirements of cloud computing data management. We also showed that these challenges clearly surpass the requirements that current data management systems are capable to address.

Based on an overview of the current state of the art in cloud computing data management, we argued that these challenges have, so far, only been addressed by using and extending relational technologies. As a consequence, we revisited the requirements of cloud computing data management and identified several opportunities for object databases. Due to the unique properties of object data models and algebras, these opportunities exist in the context of all identified requirements. Finally, these opportunities will foster technological innovation in industry and, at the same time, present interesting challenges for research in the domain of object databases. We believe that, in order for cloud computing data management to be successful, it is essential to pursue both of these directions.

To conclude, we would like to clearly state that object databases are not the only technology that needs to be considered for cloud computing data management. Rather, we have made the case that object databases have a lot to offer in the context of cloud computing. As many of their concepts align well with both the cloud computing stack and novel processing paradigms, object databases are a good basis for the integration of these other technologies.

Acknowledgment

The author would like to thank Moira C. Norrie, David Maier and Alan Dearle for the discussions about the work presented in this paper and their valuable feedback on initial drafts.

References

1. Stonebraker, M., Çetintemel, U.: One Size Fits All: An Idea Whose Time Has Come and Gone. In: Proc. Intl. Conf. on Data Engineering, pp. 2–11 (2005)
2. Abadi, D.J.: Data Management in the Cloud: Limitations and Opportunities. IEEE Data Eng. Bull. 32(1), 3–12 (2009)
3. Gounaris, A.: A Vision for Next Generation Query Processors and an Associated Research Agenda. In: Proc. Intl. Conf. on Data Management in Grid and Peer-to-Peer Systems, pp. 1–11 (2009)
4. Vesset, D.: Worldwide Data Warehousing Tools 2005 Vendor Shares. Technical Report 203229, IDC (August 2005)
5. Stonebraker, M., Abadi, D.J., DeWitt, D.J., Madden, S., Paulson, E., Pavlo, A., Rasin, A.: MapReduce and Parallel DBMS: Friends or Foes? Commun. ACM 53(1), 64–71 (2010)
6. Stonebraker, M.: The Case for Shared Nothing. IEEE Data Eng. Bull. 9(1), 4–9 (1986)
7. Golab, L., Özsu, M.T.: Issues in Data Stream Management. SIGMOD Rec. 32, 5–14 (2003)
8. Braga, D., Ceri, S., Daniel, F., Martinenghi, D.: Optimization of Multi-Domain Queries on the Web. In: Proc. Intl. Conf. on Very Large Databases, Auckland, New Zealand, August 23-28, pp. 562–573 (2008)
9. Dean, J., Ghemawat, S.: MapReduce: Simplified Data Processing on Large Clusters. In: Proc. Symp. on Operating Systems Design and Implementation, pp. 137–149 (2004)
10. DeWitt, D.J., Gray, J.: Parallel Database Systems: The Future of High Performance Database Systems. ACM Commun. 35(6), 85–98 (1992)

11. Özsu, M.T., Valduriez, P.: Principles of Distributed Database Systems, 2nd edn. Prentice-Hall, Englewood Cliffs (1999)
12. Kossmann, D.: The State of the Art in Distributed Query Processing. ACM Comput. Surv. 32(4), 422–469 (2000)
13. Kephart, J.O., Chess, D.M.: The Vision of Autonomic Computing. Computer 36(1), 41–50 (2003)
14. Gounaris, A., Smith, J., Paton, N.W., Sakellariou, R., Fernandes, A.A., Watson, P.: Adaptive Workload Allocation in Query Processing in Autonomous Heterogeneous Environments. Distrib. Parallel Databases 25(3), 125–164 (2009)
15. Deshpande, A., Ives, Z., Raman, V.: Adaptive Query Processing. Found. Trends Databases 1(1), 1–140 (2007)
16. Avnur, R., Hellerstein, J.M.: Eddies: Continuously Adaptive Query Processing. In: Proc. ACM SIGMOD Intl. Conf. on Management of Data, pp. 261–272 (2000)
17. Luo, G., Ellmann, C.J., Haas, P.J., Naughton, J.F.: A Scalable Hash Ripple Join Algorithm. In: Proc. ACM SIGMOD Intl. Conf. on Management of Data, pp. 252–262 (2002)
18. Abadi, D.J., Ahmad, Y., Balazinska, M., Çetintemel, U., Cherniack, M., Hwang, J.H., Lindner, W., Maskey, A.S., Rasin, A., Ryvkina, E., Tatbul, N., Xing, Y., Zdonik, S.: The Design of the Borealis Stream Processing Engine. In: Proc. Intl. Conf. on Innovative Data Systems Research, Asilomar, CA, USA, January 4-7, pp. 277–289 (2005)
19. Srivastava, U., Munagala, K., Widom, J., Motwani, R.: Query Optimization over Web Services. In: Proc. Intl. Conf. on Very Large Data Bases, pp. 355–366 (2006)
20. Gripay, Y., Laforest, F., Petit, J.M.: A Simple (Yet Powerful) Algebra for Pervasive Environments. In: Proc. Intl. Conf. on Extending Database Technology, 359–370 (2010)
21. Ceri, S., Brambilla, M. (eds.): Search Computing – Challenges and Directions. Springer, Heidelberg (2010)
22. Pacitti, E., Valduriez, P., Mattoso, M.: Grid Data Management: Open Problems and New Issues. J. Grid Comput. 5(3), 273–281 (2007)
23. Antonioletti, M., Atkinson, M.P., Baxter, R., Borley, A., Hong, N.P.C., Collins, B., Hardman, N., Hume, A.C., Knox, A., Jackson, M., Krause, A., Laws, S., Magowan, J., Paton, N.W., Pearson, D., Sugden, T., Watson, P., Westhead, M.: The Design and Implementation of Grid Database Services in OGSA-DAI: Research Articles. Concurr. Comput.: Pract. Exper. 17(2-4), 357–376 (2005)
24. Lynden, S., Mukherjee, A., Hume, A.C., Fernandes, A.A.A., Paton, N.W., Sakellariou, R., Watson, P.: The Design and Implementation of OGSA-DQP: A Service-Based Distributed Query Processor. Future Gener. Comput. Syst. 25(3), 224–236 (2009)
25. Olston, C., Reed, B., Srivastava, U., Kumar, R., Tomkins, A.: Pig Latin: A Not-So-Foreign Language for Data Processing. In: Proc. ACM SIGMOD Intl. Conf. on Management of Data, 1099–1110 (2008)
26. Yu, Y., Isard, M., Fetterly, D., Budiu, M., Erlingsson, Ú., Gunda, P.K., Currey, J.: DryadLINQ: A System for General-Purpose Distributed Data-Parallel Computing Using a High-Level Language. In: Proc. Symp. on Operating Systems Design and Implementation, pp. 1–14 (2008)
27. Chaiken, R., Jenkins, B., Larson, P.Å., Ramsey, B., Skakib, D., Weaver, S., Zhou, J.: SCOPE: Easy and Efficient Parallel Processing of Massive Data Sets. In: Proc. Intl. Conf. on Very Large Databases, pp. 1265–1276 (2008)

28. Abouzeid, A., Bajda-Pawlikowski, K., Abadi, D.J., Rasin, A., Silberschatz, A.: HadoopDB: An Architectural Hybrid of MapReduce and DBMS Technologies for Analytical Workloads. In: Proc. Intl. Conf. on Very Large Databases, pp. 922–933 (2009)

29. DeWitt, D.J., Paulson, E., Robinson, E., Naughton, J., Royalty, J., Shankar, S., Krioukov, A.: Clustera: An Integrated Computation and Data Management System. In: Proc. Intl. Conf. on Very Large Databases, pp. 28–41 (2008)

30. Pavlo, A., Paulson, E., Rasin, A., Abadi, D.J., DeWitt, D.J., Madden, S., Stonebraker, M.: A Comparison of Approaches to Large-Scale Data Analysis. In: Proc. ACM SIGMOD Intl. Conf. on Management of Data, pp. 165–178 (2009)

31. Atkinson, M.P., Bancilhon, F., DeWitt, D.J., Dittrich, K.R., Maier, D., Zdonik, S.B.: The Object-Oriented Database System Manifesto. In: Building an Object-Oriented Database System, The Story of O2, pp. 3–20. Morgan Kaufmann, San Francisco (1992)

32. Fegaras, L., Maier, D.: Towards an Effective Calculus for Object Query Languages. In: Proc. ACM SIGMOD Intl. Conf. on Management of Data, pp. 47–58 (1995)

33. Özsu, M.T., Blakeley, J.A.: Query Processing in Object-Oriented Database Systems. In: Modern Database Systems: The Object Model, Interoperability, and Beyond, pp. 146–174. ACM Press/Addison-Wesley Publishing Co. (1995)

34. Wang, Q., Maier, D., Shapiro, L.D.: The Hybrid Technique for Reference Materialization in Object Query Processing. In: Proc. Intl. Symp. on Database Engineering and Applications, pp. 37–46 (2000)

35. Norrie, M.C., Palinginis, A., Würgler, A.: OMS Connect: Supporting Multidatabase and Mobile Working through Database Connectivity. In: Proc. Intl. Conf. on Cooperative Information Systems, pp. 232–240 (1998)

36. Kuno, H.A., Ra, Y.G., Rudensteiner, E.A.: The Object-Slicing Technique: A Flexible Object Representation and its Evaluation. Technical Report CSE-TR-241-95, University of Michigan (1995)

37. Karlapalem, K., Li, Q.: A Framework for Class Partitioning in Object-Oriented Databases. Distrib. Parallel Databases 8(3), 333–366 (2000)

38. Lieuwen, D.F., DeWitt, D.J., Mahta, M.: Parallel Pointer-based Join Techniques for Object-Oriented Database. In: Proc. Intl. Conf. on Parallel and Distributed Information Systems, pp. 172–181 (1993)

39. Witt, D.J.D., Naughton, J.F., Shafer, J.C., Venkataraman, S.: Parallelizing OODBMS Traversals: A Performance Evaluation. The VLDB Journal 5(1), 3–18 (1996)

40. Cattell, R.G.G., Barry, D.K., Berler, M., Eastman, J., Jordan, D., Russell, C., Schadow, O., Stanienda, T., Velez, F. (eds.): The Object Data Standard: ODMG 3. Morgan Kaufmann, San Francisco (2000)

41. Norrie, M.C.: An Extended Entity-Relationship Approach to Data Management in Object-Oriented Systems. In: Proc. Intl. Conf. on the Entity-Relationship Approach, pp. 390–401 (1993)

42. Maier, D.: Representing Database Programs as Objects. In: Proc. Intl. Workshop on Database Programming Languages, pp. 377–386 (1987)

43. Dearle, A., Kirby, G.N.C., Morrison, R.: Orthogonal Persistence Revisited. In: Proc. Intl. Conf. on Object Databases, pp. 1–23 (2009)

Query Optimization by Result Caching in the Stack-Based Approach

Piotr Cybula[1] and Kazimierz Subieta[2,3]

[1] Institute of Mathematics and Computer Science, University of Lodz, Lodz, Poland
cybula@math.uni.lodz.pl
[2] Polish-Japanese Institute of Information Technology, Warsaw, Poland
subieta@pjwstk.edu.pl
[3] Institute of Computer Science, Polish Academy of Sciences, Warsaw, Poland

Abstract. We present a new approach to optimization of query languages using cached results of previously evaluated queries. It is based on the stack-based approach (SBA) and object-oriented query language SBQL. SBA assumes description of semantics in the form of abstract implementation of query/programming language constructs. Pragmatic universality of SBQL and its precise, formal operational semantics make it possible to investigate various crucial issues related to this kind of optimization. Two main issues are: organization of the cache enabling fast retrieval of cached queries and development of fast methods to recognize consistency of queries and incremental altering of cached query results after database updates. This paper is focused on the first issue concerning optimal, fast and transparent utilization of the result cache, involving methods of query normalization enabling higher reuse of cached queries with preservation of original query semantics and decomposition of complex queries into smaller ones. We present experimental results of the optimization that demonstrate the effectiveness of our technique.

1 Introduction

Caching results of previously evaluated queries seems to be an obvious method of query optimization. It assumes that there is a relatively high probability that the same query will be issued again by the same or another application, thus instead of evaluating the query the cached result can be reused. There are many cases when such an optimization strategy makes a sense. This concerns the environments where data are not updated or are updated not frequently (say, one update for 100 retrieval operations). Examples are data warehouses (OLAP applications), various kinds of archives, operational databases, knowledge bases, decision support systems, etc.

Besides the frequency of database updates, which is critical to such methods, another critical factor concerns the probability of caching query reuse. For instance, in a typical internet shop we can estimate that 90% of requests concerns some 10% of products, hence queries addressing these 10% products are worth to be cached. Such caching (in the form of HTML pages or XML files) is assumed in many commercial Web applications.

A. Dearle and R.V. Zicari (Eds.): ICOODB 2010, LNCS 6348, pp. 40–54, 2010.

Conceptually, the cache can be understood as a two-column table, where one column contains cached queries in some internal format (e.g. normalized syntactic query trees), and the second column contains query results. A query result can be stored as a collection of OIDs, but for special purposes can also be stored e.g. as an XML file enabling further quick reuse in Web applications. A cached query can be created as a side effect of normal evaluation of user query or by the database administrator in advance.

A transparency is the most essential property of a cached query. It implies that programmers need not to involve explicit operations on cached results into an application program. Caching of query results yields the major improvement in query evaluation performance, i.e. significantly decreases the response time from a database management system. The main reason is that receiving a result from a cache for a previously performed query, instead of its consecutive reevaluation, is much quicker than time-consuming query processing. In contrast to other query optimization methods, which strongly depend on the semantics of a particular query, the query caching method is independent of a query type, its complexity and a current database state.

On the other hand there are some costs of result materialization. Firstly, some memory resources are necessary for the cache storing queries and their results. Secondly, the optimization method needs some time for storing queries and their results, together with proper structures for maintenance purposes, recognizing the usability of currently cached results for new queries, removing some rarely used cached queries in order to optimal cache utilization and finally, updating cached results after database changes.

Our concept of cached queries follows the work presented in [20] and [24], devoted to a kind of a network database model. In comparison, object-oriented and XML-oriented data models and their query languages present new qualities, thus the methods that we discuss and propose are significantly different. Our research is done within the Stack-Based Approach (SBA) to object-oriented query/programming languages. SBA is a formal theory and a universal conceptual frame addressing this kind of languages, thus it allows precise reasoning concerning various aspects of cached queries, in particular, query semantics, query decomposition, query indexing in the cache, and so on. We have implemented the caching methods as a part of the optimizer developed for the query language SBQL (Stack-Based Query Language) in our last project ODRA (Object Database for Rapid Application development) devoted to Web and grid applications [18]. In [8] we have described how query caching can be used to enhance performance of applications operating on grids.

There are two key aspects concerning the development of database query optimization using cached queries:

- organization of the cache enabling fast retrieval of cached queries (for optimal queries selection and rewriting new queries with use of cached results) and optimal, fast and transparent utilization of the cache, involving methods of query normalization with preservation of original query semantics (enabling higher reuse of cached queries for semantically equivalent but syntactically different queries), decomposition of complex queries into smaller ones and maintenance of assigned resources by removing rarely used results;

- development of fast methods to recognize consistency of queries and automatic incremental altering of cached query results after database updates (sometimes re-moving or re-calculating).

In this paper we deal mainly with the first issue of the optimization method. The second aspect is widely researched in [7, 9, 10]. The paper is organized as follows. Section 2 discusses known solutions that are related to the contributions of the paper. In section 3 we briefly present the Stack-Based Approach. Section 4 describes the architecture of the caching query optimizer. Section 5 contains the description of optimization strategies, in particular query normalization, decomposition and rewrit-ing rules. Section 6 presents experimental results and Section 7 concludes.

2 Related Work

Cached queries remind materialized views, which are also snapshots on database states and are used for enhancing information retrieval [2, 4, 5]. Such materialized views are currently implemented in popular relational database systems as DB2 and Oracle [12, 19]. Materialization of query results in object-oriented algebras in the form of materialized views is considered in [1] and [13]. Some solutions for view result caching at client-side in object and relational databases and for optimal combi-nation of materialized results in cache to answer a given query are presented in [11]. In [6] and [21] a solution for XML query processing using materialized XQuery views is proposed. Authors present an algebraic approach for incremental update of materialized XQuery views built using some selected query operators.

There are, however, two essential differences between materialized views and cached queries. The first one concerns the scale. One can expect that there will be at most dozens of materialized views, but the number of cached queries could be thou-sands or millions. Such scale difference implies the conceptual difference. The second difference concerns transparency: while materialized views are explicit for software developers, cached queries are an internal feature that is fully transparent for them. Our research is just about how this transparent mechanism can be used to query opti-mization, assuming no changes to syntax, semantics and pragmatics of the query language itself.

Cached queries are also similar to database indices [14]. Both cached queries and indices are server-side auxiliary structures that are used for fast retrieval. For instance, an item of a dense index having a key value "designer" from the job attribute of the Emp table can be perceived as a cached SBQL query

<Emp **where** job = "designer", collection of OIDs>

where the 'collection of OIDs' is a non-key index value (object identifiers returned by the query) associated with the key value "designer" in the index. For fast retrieval such a structure can be implemented e.g. as a hash table or B-tree, similarly to the methods of organizing indices. However cached queries are conceptually different from indices. Indices usually materialize very simple queries, while in general we can cache arbitrarily complex queries if they are promising to be reused. For instance, we can consider caching queries containing multi-parameterized selections, aggregations, long path expressions, grouping, etc. Indices are also made in advance, while cached queries are a side effect of previous query evaluations. Indices contain items for all database values of the given attribute (e.g. for all values of the job attribute), while cached queries usually contain only a subset of them. For these reasons cached queries imply quite new research and implementation problems.

New Oracle 11g database system [19] offers also caching of SQL and PL/SQL results. The cached results of SQL queries and PL/SQL functions are automatically reused while subsequent invocation and updated after database modifications. On the other hand, in opposition to our proposal, materialization of the results is not fully transparent. Query results are cached only when query code contains a comment with a special parameter 'result_cache', so the evaluation of old codes without the parameter is not optimized.

Query cache is also implemented in MySQL database [17], where only full SELECT query texts together with the corresponding results are stored in the cache. In the solution caching does not work for subselects and stored procedure calls (even if it simply performs a SELECT query). Queries must be absolutely the same - they have to match byte by byte for cache utilization, because of matching of not normalized query texts (e.g. the use of different letter case causes insertion of different queries into the query cache).

There is in Microsoft .NET query language LINQ [15] some kind of query result caching as an optimization technique for often requested queries, but it is also not transparent for programmers. They have to explicitly place the results of queries into a list or an array (calling one of the methods ToList or ToArray) and in a consequence each subsequent request of such query will cause getting its results from the cache instead of the query reevaluation.

But there is not any result caching solutions implemented in current leading commercial and non-commercial object-oriented database systems. Most of them bases their query languages on OQL (Object Query Language) proposed as a model query language by ODMG (Open Database Management Group) [3]. Only a cache of objects is introduced in some implementations for fast access of data in a distributed database environment.

3   Overview of the Stack-Based Approach (SBA)

The *Stack-Based Approach* (SBA) along with its query language SBQL (*Stack-Based Query Language*) is the result of investigations into a uniform conceptual and semantic platform for integrated query and programming languages for object-oriented databases. SBA assumes that query languages are a special case of programming languages. The approach is abstract and universal, which makes it relevant to a general object model. In SBA each object has the following features:

- *internal identifier* ($i \in I$) – OID, which cannot be directly written in queries,
- *external name* ($n \in N$) that is used to access objects from queries,
- *content* ($v \in V$) that can be a value (integer, string, blob, etc.), a link, or a set of objects. Atomic values include also codes of procedures, functions, methods, views, etc.

Formally, objects are recursively modeled as triples, where i, i_1, $i_2 \in I$, $n \in N$, $v \in V$: *atomic objects* as $<i, n, v>$, *link objects* as $<i_1, n, i_2>$, *complex objects* as $<i, n, S>$, where S denotes a set of objects.

SBA respects the naming-scoping-binding principle, which means that each name occurring in a query is bound to the appropriate run-time entity (an object, attribute, method parameter, etc) according to the scope of this name. One of its basic mechanisms is an environment stack (ENVS), which is responsible for scope control and for binding names. In contrast to classical stacks, it does not store objects, but some structures built upon object identifiers, names, and values. SBA assumes the principles of semantic relativity, orthogonal persistence and full internal identification of runtime entities (objects).

Stack-Based Query Language (SBQL) is thoroughly described in [22, 23]. The language has several implementations - for the XML DOM model, for OODBMS Objectivity/DB, and recently for the object-oriented ODRA system [18]. SBQL is based on an abstract syntax and the principle of *compositionality*: it avoids syntactic sugar and syntactically separates as far as possible query operators. In contrast to SQL and OQL, SBQL queries have the useful property: they can be easily decomposed into subqueries, down to atomic ones, connected by unary or binary operators. The property simplifies implementation and greatly supports query optimization. The SBQL operational semantics introduces another stack, known as query result stack (QRES), for storing temporary and final query results. The two stacks architecture is the core of the Stack-Based Approach (SBA).

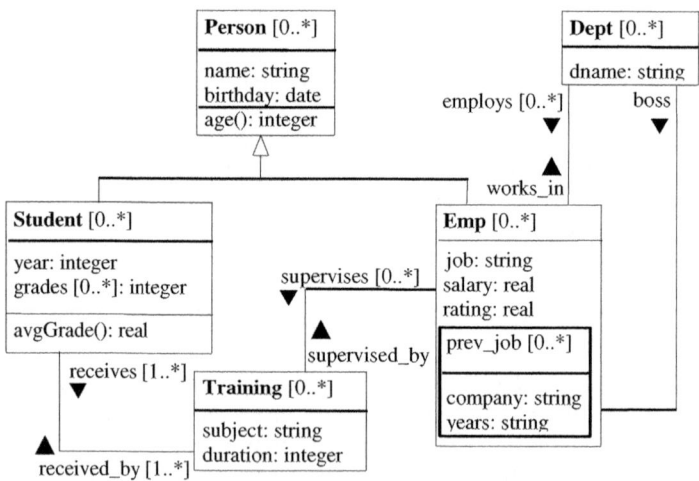

Fig. 1. Class diagram of the example database

The syntax of SBQL is as follows:

- A single name or a single literal is an (atomic) query. For instance, *Student*, *name*, *year*, *x*, *y*, "Smith", 2, 2500, etc, are queries.
- If q is a query, and σ is a unary operator (e.g. sum, count, distinct, sin, sqrt), then $\sigma(q)$ is a query.

- If q_1 and q_2 are queries, and θ is a binary operator (e.g. **where**, dot, **join**, +, =, **and**), then $q_1 \, \theta \, q_2$ is a query.
- There are not other queries in SBQL.

SBQL, unlike SQL and other query languages, avoids big syntactic and semantic patterns. Atomic queries are single names and literals. Nested queries can be arbitrarily composed from atomic and nested queries by unary and binary operators, providing they have a sense for the programmer and do not violate typing constraints. Classical query operators, such as selection, projection/navigation, join, quantifiers, etc. are also binary operators, but their semantics involves ENVS. For this reason they are called "non-algebraic" – their semantics cannot be expressed by any algebra designed in the style of the relational algebra. Below we present the exemplary operational semantics for one of the often used "non-algebraic" operator of projection (dot operator):

1. Initialize an empty bag (*eres*).
2. Execute the left subquery.
3. Take a result collection from QRES (*colres*).
4. For each element *el* of the *colres* result do:
    4.1. Open new section on ENVS.
    4.2. Execute function *nested(el)*.
    4.3. Execute the right subquery.
    4.4. Take its result from QRES (*elres*).
    4.5. Insert *elres* result into *eres*.
5. Push *eres* on QRES.

Step 4.2 employs a special function *nested* which formalizes all cases that require pushing new sections on the ENVS, particularly the concept of pushing the interior of an object. This function takes any query result as a parameter and returns a set of binders. For the operator **where** all steps are the same except for 4.5 and a new 4.6:
    4.5. Verify whether *elres* is a single result (if not exception is raised).
    4.6. If *elres* is equal to *true* add *el* to *eres*.
For the navigational join operator (**join**) the steps are:
    4.5. Perform Cartesian Product operation on *el* and *elres*.
    4.6. Insert obtained structure into *eres*.

For SBA optimization examples presented in next sections we assume the class diagram in Fig. 1. The schema defines five classes (i.e. five collections of objects): *Training*, *Student*, *Emp*, *Person*, and *Dept*. The classes *Training*, *Student*, *Emp* and *Dept* model students receiving trainings, which are supervised by employees of departments organizing these trainings. *Person* is the superclass of the classes *Student* and *Emp*. *Emp* objects can contain multiple complex *prev_job* subobjects (previous jobs). Names of classes (as well as names of attributes and links) are followed by cardinality numbers, unless the cardinality is 1.

4   Query Optimizer Architecture

In most commercial client/server database systems (c.f. SQL processors) all the query processing is performed on the server. In SBA majority of query processing is shifted to the client side, to avoid server overloading. Fig. 2 presents query processing architecture in SBA. Firstly, similarly to indices, the *query cache registry* is stored at the server. The reason is the ability to share cached results between all users, and easier way to maintain the results after data modifications. Hence the client-side query optimizer looks up in this registry before starts optimization and processing a given query. Secondly, in opposite to the traditional approaches, the storing of the query result should also be processed differently. Because only the client knows the form of the query and its result, the client is responsible to send the pair *<query, result>* to the server in order to include it within the query cache registry (as presented in the next section).

One of the main components of the optimization uses cached results is query cache registry. Since the amount of cached queries can be very large, the structures used to implement the query registry must ensure very fast access and search capabilities. We propose a linear hashing table [16] with a single, primary key as fast and efficient search data structure for cached results. The single key retrieval is very simple to implement and is independent of the query type - the response time is short and always the same regardless of the complexity of request. There are several candidate solutions for the search key. The simplest one is simply a query text (normalized using some sophisticated techniques mentioned in the next section) considered as a character string. Taking into account the equivalence of text of a query and its syntactic tree, instead of difficult searching within the set of syntactic trees of cached queries, we can search in the efficient and proved linear hashing index structure containing texts. Non-key values of the query index are references to the *metabase* nodes (MB_ID) containing meta-information concerning cached queries:

- signature of cached results for type-checking purposes;
- reference to the object store node (DB_ID) containing compiled cached query (for further reevaluation), cached results, statistic data and auxiliary structures for efficient update of the results after database change.

Queries are cached both in the physical object store DB (persistent memory cache guarantying maintenance of the cache after database restart) and in a virtual object store TMP (a volatile main memory cache guarantying fast access). Query optimization using cached results involves main subsystems of query evaluation environment, such query optimizer, query interpreter and query cache registry placed in the object store and a metabase.

We propose the following scenario of the optimization using cached queries in query evaluation environment for SBA (step numbers correspond to subsequent label numbers in Fig. 2):

1. A user sends a query to a client-side database interface.
2. The *parser* receives it and transforms into a *syntactic tree*.
3. The tree is statically evaluated for *type checking* with the use of the static stacks (ENVS and QRES) and a database schema stored in the metabase at the server-side. After successful static evaluation the nodes of the query tree are augmented with type signatures for easier optimization reasoning.
4. The tree is sent to the *cache optimizer* being one in a sequence of optimizers employed at the client-side database system.

5. The cache optimizer rewrites it using strategies presented in the next section. It employs the server-side *cache manager* which proposes optimal matching of results cached in the query cache registry, performs proper steps for a new query caching if suggested by the optimizer and maintains cache usage statistics for optimal cache utilization and cleaning. For each new cached query the manager generates additional structures, which describe a subset of involved objects for maintenance purposes. The system updates cached results after changes in the database accordingly to the algorithms presented thoroughly in [7, 9, 10].

6. Finally a modified, optimized, type-checked and compiled *query evaluation plan* is produced and sent for execution by the *query interpreter*.

7. The plan is evaluated by the query interpreter. Some parts of the plan rewritten by the cache optimizer suggest taking the cached results from the server-side object store instead of reevaluation of them. For new queries being candidates for caching the interpreter generates their results and sends it to the cache manager for storing at the database server. The operations are described in the next section.

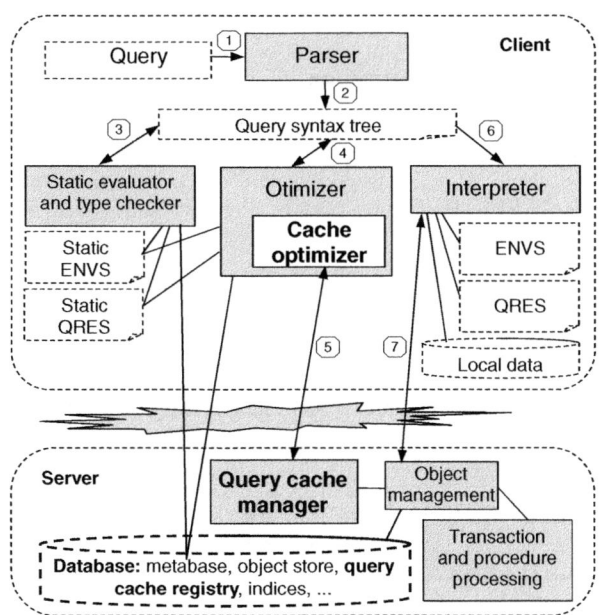

Fig. 2. Query optimization steps

5   Query Optimization Using Cached Results

The cache optimizer acts (step 5 in Fig. 2) in four steps. (1) The optimizer normalizes a query, then (2) the query is decomposed. (3) The main step – the query is analyzed and rewritten, and finally (4) it is type-checked before forwarding to other optimizers or final compilation into an evaluation plan. In the last two steps the optimizer communicates with the server-side cache manager. Below we present a short description of strategies developed in all the optimization steps. Precise algorithms for each solution are included in [7].

5.1 Query Normalization

To prevent from placing in the cache queries with different textual forms but the same semantic meaning we introduce several query text *normalization methods*. These methods are applied in a way of reconstructing a query text from early generated query syntactic tree or directly by change some nodes or their order within the tree. The main methods are:

- **alphabetical ordering of operands for operators**, which for a succession of operands is not substantial, such as comparing operators (=, ≠, <=, <, >, >=), arithmetic operators (+, *), logical operators (**or**, **and**), operators of sum and intersection of sets, i.e.:

  ```
  Emp where salary >= 1100 or salary = 1000
  ```

  is normalized to:

  ```
  Emp where 1000 = salary or 1100 <= salary
  ```

- **ordering of operators** (e.g. putting sum operations before subtractions or multiplication before division), i.e.:

  ```
  a / b / c * d / e
  ```

  is transformed to:

  ```
  a * d / b / c / e
  ```

- **unification of auxiliary names** used by the programmer for **as** or **group as** operator, but only if such an operator doesn't finalize the evaluation of the query (it is not the root of the syntactic tree, which case is easy to recognize based on query result signature evaluated earlier by the static evaluator), i.e.:

  ```
  (((Emp where salary > 1000) as e)
     join (e.works_in.Dept as d)).
     (e.name,d.dname)
  ```

  is normalized to:

  ```
  (((Emp where salary > 1000) as $cache_aux1)
     join ($cache_aux1.works_in.Dept as $cache_aux2)).
     ($cache_aux1.name,$cache_aux2.dname)
  ```

5.2 Query Decomposition and Rewriting

After normalization phase query is virtually decomposed, if possible, into one or many simpler candidate subqueries. *Query decomposition* is a useful mechanism to speed up evaluating a greater number of new queries. If we materialize a small independent subquery instead of a whole complex query, then the probability of reusing of its results is risen. In addition, a simple semantic of the decomposed query reduces the costs of its updating. Each, isolated while decomposition process, subquery and finally a whole query is independently analyzed and rewritten in context of the set of cached queries defined in the query cache registry and if it hasn't yet cached, it becomes a new candidate for caching.

Too simple queries (without object names or non-algebraic operators) are omitted. While analyzing, query is converted to the text form and the optimizer performs search process using the query index stored in the query cache registry. If found, the tree of the query is replaced with a call of a special *cache function* parameterized with unique references to nodes of matched cached query in the metabase (MB_ID) and the object store (DB_ID) – these parameters are non-key elements of cached query index mentioned earlier).

Each not yet cached candidate query is also replaced with a call of the cache function – new cached query is placed into the query index with its new created meta and data node references (MB_ID and DB_ID). In this case a new query node in the object store (identified by DB_ID) doesn't contain query results – it is marked as "not fully cached" and will be populated with its results while the first need of use (when the interpreter will evaluate it). A new meta node of the query (identified by MB_ID) contains DB_ID and a type signature of the candidate query results (evaluated with use of static query evaluator).

We use such decomposition techniques as:

- **factoring out independent subqueries** (being thoroughly investigated in [22]) – instead of caching such a complex query as:

```
Emp where salary < ((Emp where name = "Smith").salary)
```

we isolate an internal independent query:

```
(Emp where name = "Smith").salary
```

which is matched or proposed as cached query uniquely identified by its node references MB_ID and DB_ID, and finally the original query is rewritten to:

```
Emp where salary < $cache_fun(MB_ID,DB_ID)
```

The produced query can be then cached or not.

- **removing path expressions** finalizing query evaluation, that is, isolating a query without such expressions only if they are quickly evaluable (thanks to referential nature of object-oriented database). If a query is finalized with a sequence of navigational operators (dot) or the constructor of a structure (**struct**) containing such sequences, and all the objects within such expressions are unique subobjects or reference objects (with cardinality 1 or 0..1), the longest expressions fulfilling this condition are cut forming simpler independent query for caching, i.e. query:

```
(Training where count(received_by) > 12).(subject,
duration, supervised_by.Emp.salary)
```

is ended with an implicit structure constructor with a path expression. Each expression has the cardinality 1, so all expressions (and in consequence structure constructor, too) are ignored while isolating the query:

```
(Training where count(received_by) > 12)
```

and finally rewriting the input query to:

```
$cache_fun(MB_ID,DB_ID).(subject, duration,
supervised_by.Emp.salary)
```

In case of another query:

```
(Dept where dname = "Database").employs.Emp.prev_job
```

both *prev_job* (subobject) and *employs* (reference object) attributes have cardinality 0..*, so the optimal solution is to cache the whole query.

- **factoring out aggregations** (avg, min, max, sum, count), which are in many cases time consuming queries and can be interpreted as virtual materialized attributes of database objects, i.e. queries containing **join** operator, such as:

```
Dept join avg(employs.Emp.salary)
```

are cached as a group of cached queries for each *Dept* object instance which becomes an additional third parameter of cache function *$cache_fun*.

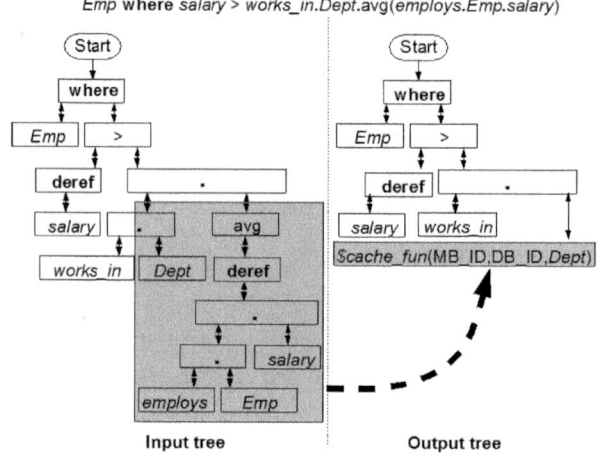

Fig. 3. Sample query optimization

In such a case another query:

```
Emp where salary >works_in.Dept.avg(employs.Emp.salary)
```

is decomposed by isolating cached query:

```
Dept.avg(employs.Emp.salary)
```

and rewriting the whole query as follows:

```
Emp where salary >works_in.$cache_fun(MB_ID,DB_ID,Dept)
```

This optimization case on a query syntactic tree is presented in Fig. 3.

- **transforming queries into equivalent forms** using operations on Boolean expressions and on sets of query results (bags) – thanks to the distributivity property of the selection operator in SBQL (**where**) it is possible to decompose queries with complex predicates containing some logical operators (**or**, **and**, **not**) into two or more simpler queries joined by set operators (sum, intersection, subtraction). For instance, the complex query:

```
Emp where (job = "clerk") or (job = "consultant")
```

is transformed into query:

```
(Emp where job = "clerk") ∪ (Emp where job =
"consultant")
```

and finally into:

```
$cache_fun(MB_ID1,DB_ID1) ∪ $cache_fun(MB_ID2,DB_ID2)
```

5.3 Usage of Cached Results and Cache Adaptability

After all optimization steps the compiled evaluation plan is executed by the interpreter at the client-side (step 7 in Fig. 2). Call of the special cache function (with MB_ID and DB_ID parameters) inserted by the optimizer causes requesting for materialized results of used cached query stored in the query cache registry at the database server. Not yet cached ("not fully cached") or marked for update queries are evaluated, their results are sent to the registry (where placed in proper node identified by given DB_ID) and immediately utilized. Results of some queries are not placed in the cache, as a result of cache policies configured by system administrator:

- minimal query evaluation time,
- minimal query usage count for placing into the cache,
- maximal size of cached results for one query.

Queries not fulfilling the conditions will be normally evaluated each time they will be requested or until their features change.

If the result of a cached query is requested, the cache manager updates its use counter placed in the object store node of the query. Use counters are used to generate *global cache statistics* implemented as priority lists of use levels in the form of MRU lists. Each level is treated as a range of query usage. Sample levels are 1..2, 3..5, 5..20, 21..1000, 1001..∞. Each level is assigned with:

- the number of cached queries which usage counters belong to the level's range;
- the total size of results of all cached queries of the level.

After the use of some cached query, its usage counter is incremented and as a result the query may move to higher usage level. After query reevaluation (while first use, too) or update of its results, the size of cached results may change, so the total size of appropriate usage level changes too. The query cache manager controls the cache by deleting unprofitable, rarely used cached queries (using the statistics) or queries too often updated. Such cache adaptability is performed under the control of the administrator, who configures cache system parameters, such as:

- maximal total cache size,
- maximal percent of the cache utilization (usage size),
- maximal percent of the cache utilization after cache reduction (clear size).

After each change of cached results (insertion or update), the cache manager recalculates total sizes of each changed use levels, summarizes all levels, and if the usage size permitted was exceeded, calculates a reduction factor being the top constraint of the lowest use level, for which summary sizes for all levels below are sufficient to reduce the cache size to the proposed clear size. All cached queries (with their results and all data stored in the metabase and the object store) are deleted beginning from the queries belonging to the lowest use levels until the level with the reduction factor calculated. Appropriate statistics for cleared use levels are also cleared.

Optionally, use counters and statistics for all remaining, still cached queries, are set to zero, as a way to give equal rights to all other queries, especially those of new cached queries that will be often utilized in the near future but not in the past.

6 Experimental Results

We have tested the performance of the optimizer by calculating response times for 100 subsequent requests using a set of the same queries retrieving data from database containing over 100000 objects being instances of *Dept* or *Emp* class according to the schema presented in Fig. 1.

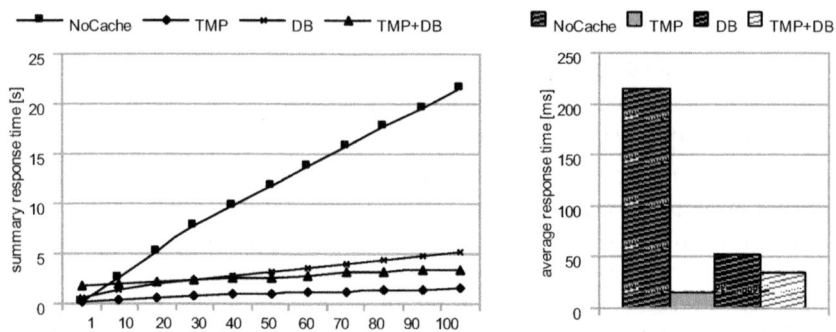

Fig. 4. Efficiency of optimization using cached queries

Some input queries with the same semantics were syntactically different, but after the normalization or decomposition they became unified. We have compared four optimization strategies: without optimization (NoCache), caching in volatile memory (TMP), caching in persistent memory (DB) and mixed caching (TMP+DB). The results presented in Fig. 4 show that in case of the TMP strategy average response time is more than 10 times shorter than response without using of the cache. In many cases, especially for more complex queries (using multi-parameterized predicates or aggregations), responses were 100 times faster.

7 Conclusions and Future Work

We have presented an approach to optimization of query execution using caching of the results of previously answered queries. Our solution addresses the stack-based approach to object-oriented query languages. The cached queries method as a tool for optimization ensures short and scalable response time to any user request types. Proper structures and strategies for fast retrieval of cached queries results have been proposed. We have presented organization of the query cache registry (for optimal queries selection and rewriting new queries with the use of cached results) and optimal, fast and transparent utilization of the cache. Methods of query normalization were developed, with preservation of the original query semantics (enabling higher reuse of cached queries for semantically equivalent but syntactically different queries). Query decomposition of complex queries into smaller ones was presented and the maintenance strategies for optimal managing of cache resources by removing rarely used, unprofitable results were described. Some experimental results of the optimization were introduced that demonstrate the effectiveness of our method.

The work on cached queries is continued. There are many open research areas concerning this optimization method. The main areas concern some additional features of SBA and SBQL not mentioned in this paper, such as inheritance and dynamic object roles. In general, the problem is practical rather than theoretical, hence much effort should be devoted to experiments with different strategies of caching queries and keeping in sync their stored results.

References

1. Ali, M.A., Fernandes, A.A.A., Paton, N.W.: MOVIE: An Incremental Maintenance System for Materialized Object Views. In: Proc. of Data & Knowledge Engineering, vol. 47, pp. 131–166 (2003)
2. Blakeley, J.A., Larson, P., Tompa, W.M.: Efficiently Updating Materialized Views. In: Proc. of ACM SIGMOD, pp. 61-71 (1986)
3. Cattell, R.G.G., Barry, D.K. (eds.): The Object Data Standard: ODMG 3.0. Morgan Kaufmann, San Francisco (2000)
4. Chaudhuri, S., Krishnamurthy, R., Potamianos, S., Shim, K.: Optimizing Queries with Materialized Views. In: Proc. of Intl. Conf. on Data Engineering, pp. 190-200 (1995)
5. Chen, C.M., Roussopoulos, N.: The Implementation and Performance Evaluation of the ADMS Query Optimizer: Integrating Query Result Caching and Matching. In: Proc. of Intl. Conf. On Extending Database Technology (1994)
6. Chen, L., Rundensteiner, E.A.: ACE-XQ: A CachE-ware XQuery Answering System. In: Proc. of WebDB, pp. 31-36 (2002)
7. Cybula, P.: Cached Queries as an Optimization Method in the Object-Oriented Query Language SBQL, Ph.D. Thesis, Institute of Computer Science, Polish Academy of Sciences, Warsaw (2010) (in Polish)
8. Cybula, P., Kozankiewicz, H., Stencel, K., Subieta, K.: Optimization of Distributed Queries in Grid via Caching. In: Meersman, R., Tari, Z., Herrero, P. (eds.) OTM-WS 2005. LNCS, vol. 3762, pp. 387–396. Springer, Heidelberg (2005)
9. Cybula, P., Subieta, K.: Cached Queries in the Stack-Based Approach, Institute of Computer Science, Polish Academy of Sciences, Report 985, Warsaw (2005)

10. Cybula, P., Subieta, K.: Query Optimization through Cached Queries for Object-Oriented Query Language SBQL. In: van Leeuwen, J., Muscholl, A., Peleg, D., Pokorný, J., Rumpe, B. (eds.) SOFSEM 2010. LNCS, vol. 5901, pp. 308–320. Springer, Heidelberg (2010)

11. Dar, S., Franklin, M.J., Jonsson, B.T., Srivastava, D., Tan, M.: Semantic Data Caching and Replacement. In: Proc. of VLDB (1996)

12. IBM DB2 Universal Database SQL Reference, Version 8, vol. 2, (2002),Faster Federated Queries with MQTs. DB2 Magazine 8(3) (2003)

13. Kemper, A., Moerkotte, G.: Access Support in Object Bases. In: Proc. of ACM SIGMOD, pp. 364-376 (1990)

14. Kowalski, T., Wiślicki, J., Kuliberda, K., Adamus, R., Subieta, K.: Optimization by Indices in ODRA. In: Proceedings of the First International Conference on Object Databases, ICOODB 2008, pp. 97-117 (2008)

15. LINQ:NET Language-Integrated Query, Microsoft Corporation (February 2007)

16. Litwin, W.: Linear hashing: A new tool for file and table addressing. In: Proc. of 6th VLDB, pp. 212-223 (1980)

17. MySQL 5.4 Reference Manual, Chapter 7.5.5: The MySQL Query Cache (2009)

18. ODRA (Object Database for Rapid Application development), Description and Programmer Manual, http://sbql.pl/various/ODRA/ODRA_manual.html

19. Oracle 9i Materialized Views, An Oracle White Paper (May 2001), On Oracle Database 11g, Oracle Magazine XXI (5) (2007)

20. Rzeczkowski, W., Subieta, K.: Stored Queries – a Data Organization for Query Optimization. Proc. of Data & Knowledge Engineering 3, 29–48 (1988)

21. EL-Sayed, M., Wang, L., Ding, L., Rundensteiner, E.A.: An Algebraic Approach for Incremental Maintenance of Materialized XQuery Views. In: Proc. of WIDM (2002)

22. Subieta, K.: Theory and practice of object query languages. Polish-Japanese Institute of Information Technology (2004) (in Polish)

23. Subieta, K., Beeri, C., Matthes, F., Schmidt, J.W.: A Stack Based Approach to Query Languages. In: Proc. of 2nd Springer Workshops in Computing (1995)

24. Subieta, K., Rzeczkowski, W.: Query Optimization by Stored Queries. In: Proc. of VLDB, pp. 369-380 (1987)

A Flexible Object Model and Algebra for Uniform Access to Object Databases

Michael Grossniklaus[1], Alexandre de Spindler[2],
Christoph Zimmerli[2], and Moira C. Norrie[2]

[1] Dipartimento di Elettronica e Informazione, Politecnico di Milano
I-20133 Milano, Italy
grossniklaus@elet.polimi.it
[2] Institute for Information Systems, ETH Zurich
CH-8092 Zurich, Switzerland
{despindler,zimmerli,norrie}@inf.ethz.ch

Abstract. In contrast to their relational counterparts, object databases are more heterogeneous in terms of their architecture, data model and functionality. To this day, this heterogeneity poses substantial difficulties when it comes to benchmark or interoperate object databases. While standardisation proposals have been made in the past, they have had limited impact as neither industry nor research has fully adopted them. We believe that one reason for this lack of adoption is that these standards were too restrictive and thus not capable of dealing with the heterogeneity of object databases. In this paper, we propose a uniform interface for access to object databases that is based on a flexible object model and algebra.

1 Introduction

Since their emergence in the 1980s, object databases have always been heterogeneous to an extent far greater than their relational siblings. One reason for heterogeneity is the fact that object databases are situated at the intersection of database management and object-oriented systems [1]. As a consequence, different object databases provide different sets of capabilities depending on their origin. On a very general level, the two approaches can be characterised in terms of whether they aim at supporting the compile-time or the run-time of an object data management system. Typically, object-oriented systems focus on aspects related to the design and development, whereas database management systems also address issues related to operation and evolution.

This difference is most pronounced in the object data models on which these systems are based. Models originating from object-oriented systems emphasise aspects such as encapsulation and language integration [2] and, since their main goal is to persist the objects of a programming language, these data models are usually very similar to, or even tied in with, the one of the language. In contrast, models that emerged from database management are designed to support traditional database features such as concurrency and recovery through transactions

A. Dearle and R.V. Zicari (Eds.): ICOODB 2010, LNCS 6348, pp. 55–69, 2010.

and to efficiently query large object graphs. Additionally, these models tend to address issues related to the longevity of data and, therefore, provide features to support object and schema evolution such as roles and dynamic typing.

While the different origins have led to a diverse palette of systems that are all uniquely suited to address specific application requirements, they have also hindered interoperability, data exchange, performance evaluation and, as argued by Greene [3], ultimately market adoption. Early on, efforts to rectify this situation have been undertaken in terms of defining [1], benchmarking [4,5] and standardising [6] object databases. And even though these attempts have all made important contributions, they have failed to fully deliver on the hopes invested in them. Successful object databases have become so by occupying niche markets and expanding from there, rather than by following definitions and implementing standards. We believe that one reason for this lack of adoption is that the proposals were too restrictive in the sense that the trade-off between a common core and individual strengths was not well balanced.

Nevertheless, as object databases have recently gained importance in both academia and industry, it is critical to also resume these standardisation efforts. This requirement has also been identified by the Object Management Group (OMG) which recently formed a working group to develop the next-generation object database standard [7]. We believe that the current proposal is far too generic and, in this paper, propose an alternative object model and algebra that offers a better trade-off between diversity and specificity. In the context of this model, we have also defined an algebra that supports both unordered and ordered collections with or without duplicates. Based on this model and algebra, we propose an interface to provide uniform access to object databases.

We begin in Sect. 2 with the background and discussion of related work. The object data model and corresponding algebra are presented in Sect. 3 and Sect. 4, respectively. In Sect. 5, we discuss a prototype implementation of the proposed interface that serves as a proof-of-concept. The contributions of this work as well as open issues are discussed in Sect. 6 and we conclude in Sect. 7.

2   Background

Several efforts to standardise object databases in terms of object data models and algebras have been made in the past or are still ongoing. We start by summarising the most influential approaches, before introducing the background of the object representation used in our proposal.

The best-known object database standard was defined by the Object Data Management Group (ODMG) [6]. Its object data model is based on the OMG object model and distinguishes between modelling primitives with and without unique identifier, called objects and literals, respectively. An object has a state comprised by its attributes and relationships as well as behaviour given by its methods. Objects are defined by types that consist of a specification and an implementation part. The former defines the abstract state and behaviour, while the latter furnishes a concrete realisation of the specification through a language binding. Abstract types are specified in terms of interfaces that define abstract behaviour and classes that define abstract state and behaviour. For classes, the model supports only single inheritance, whereas for interfaces multiple inheritance is allowed. Finally, predefined collection types such as set, bag, list, array and dictionary are available both as objects and as literals.

Following the renewed interest in object databases, OMG recently resumed standardisation efforts and formed the object database technology working group. The proposal in the current white paper [7] is based on a Stack-Based Architecture (SBA) [8] that features a storage model and a query language. The storage model uses ⟨subject, predicate, object⟩ triples to represent objects. The formalisation of this model is straightforward and therefore its main advantage. However, we believe that the fact that it is not specific to object databases and hence does not capture their essential features makes it unsuitable as a standard. It has been shown that storages based on triples are generic to the point of being able to represent any data model [9]. As a consequence, the current proposal has to be considered a step backwards as its low level of granularity cannot compete with earlier and semantically richer models, such as OEM [10] that uses quadruples to represent objects or the previously discussed ODMG data model.

In order to interact with object data, algebras and query languages have been defined in addition to data models. The Object Query Language (OQL) [6] was defined within the ODMG standard. OQL is a declarative query language with a syntax similar to SQL. The semantics, however, is quite different as OQL operates on sets of objects and is capable of handling path expressions. Unlike the ODMG data model that is supported by some vendors, OQL has not seen widespread adoption. Today, the Versant Query Language (VQL) [11] represents the most complete implementation, even though it only supports a very limited subset of OQL. The Stack-Based Query Language (SBQL) [8] is based on an algebra that complements the stack-based architecture introduced above. SBQL queries can be expressed using its proprietary syntax or through SBQL4J, a language-integrated query interface for the Java programming language. The latter is again confirmation of the fact that there is a trend in object databases to integrate the query language with the programming language. This approach has been pioneered by Microsoft's Language-Integrated Query (LINQ) [12,13] which is capable of accessing object, relational and XML data uniformly. Other approaches that fall into this category are db4o's programmatic query interfaces [14], namely Native Query (NQ) and Simple Object Data Access (SODA). Acknowledging this development, we are convinced that a future object database standard should specify a programmatic or language-integrated query interface, rather than a stand-alone query language.

The object model and algebra that we propose as the theoretical foundation for building a standardised interface to object databases is based on object-slicing [15]. An object representation that uses the object-slicing technique is a suitable basis for a standard as it is flexible enough to capture the diversity of object databases while, at the same time, specific enough to address their unique requirements. For example, it can uniformly represent object models regardless of whether they use single or multiple inheritance and whether multiple instantiation is possible or not [16]. In the past, object-slicing has, therefore, been proposed as an implementation technique to support features such as views, schema evolution, versions and roles. MultiView [17,18] is an implementation of object-slicing on top of GemStone and has been applied both to object-oriented views and schema evolution. While MultiView implements object-slicing based on an object database, Iris [19] follows the same approach but uses a relational back-end to store its objects. This approach is similar to more recent Object-Relational Mapping (ORM) tools that also persist objects in relational databases using model mapping [20]. However, while MultiView and Iris assume a fixed mapping between classes and so-called implementation objects, Hibernate [21], for example, offers several mapping strategies to define how objects are stored.

In summary, previous work has focused on object-slicing at the implementation level to support advanced features and to store objects flexibly. In contrast, our proposal is to leverage object-slicing at the conceptual level to unify the different approaches that exist. Unlike earlier standards, our approach recognises the importance of having diverse object databases. Therefore, our main goal is not to limit these systems by forcing them to adopt a restrictive interface. On the contrary, we propose a uniform and consistent interface to object databases that could easily be implemented by existing systems. As a consequence, the focus of our interface is more on data exchange and benchmarking, rather than application development and portability.

3   Object Data Model

In this section, we present an object data model based on object-slicing [15]. Figure 1 introduces the example used to illustrate our approach. The left shows a class hierarchy, rooted at class Contact with subclasses Organisation, Person and Private. The graphical representation of two objects based on object-slicing is given on the right. Object id_1 is an instance of class Person, whereas object id_2 is an instance of class Organisation. As can be seen, both objects consist of two so-called *object slices* which we refer to as *information units*. Each information unit corresponds to exactly one class and stores the attribute values for the fields declared by that class. Object instantiation in our model is captured by the *dress* and *strip* primitives that add or remove information units, respectively. As shown in the figure, object id_1 can be instantiated with class Private using a *dress* operation, whereas object id_1 could be reclassified as an instance of Contact using a *strip* operation. Based on this representation, we now present the formal definition of the object data model.

The type system of our object data model distinguishes four different kinds of types—base, object, structured and extent types—that describe the domain $\mathbb{T}$ of all possible values $\mathbb{V}$. Let $\mathbb{T}^* = \{\mathbb{T}_{base}, \mathbb{T}_{obj}, \mathbb{T}_{struct}, \mathbb{T}_{ext}\}$, then $\forall\, \mathbb{T}_i, \mathbb{T}_j \in \mathbb{T}^* : \mathbb{T}_i \neq \mathbb{T}_j : \mathbb{T}_i \cap \mathbb{T}_j = \emptyset$ and $\mathbb{T} = \bigcup_{\mathbb{T}_i \in \mathbb{T}^*} \mathbb{T}_i$. We will describe each of these types in more detail.

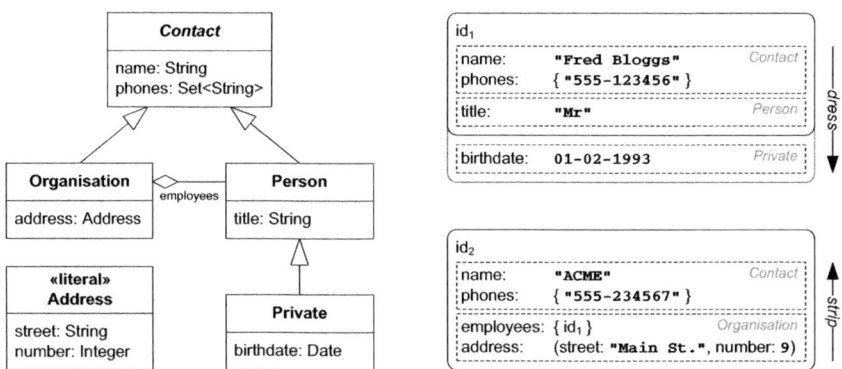

Fig. 1. Overview of our approach

Base Types. A base type $T_{base} \in \mathbb{T}_{base}$ defines the, possibly infinite, domain of a basic type that is predefined by the object database. As a consequence, $\mathbb{T}_{base}$ can change from one system to another. For the scope of this paper, we assume the following definition.

$$\mathbb{T}_{base} = \{\text{BOOLEAN}, \text{INTEGER}, \text{REAL}, \text{DATE}, \text{STRING}\}$$

We use the names of the base type as a short-hand to denote their value domains. For example, BOOLEAN is used to denote $T_{boolean} = \{\text{true}, \text{false}\}$.

A *base value*, $v_{base} \in \mathbb{V}_{base}$, has no identity and is said to be an instance of a base type $T_{base} \in \mathbb{T}_{base}$, denoted as $v_{base} \vdash T_{base}$, iff $v_{base} \in T_{base}$. Generally, base types and their instances cannot be explicitly created, modified or deleted as their existence is taken for granted.

Object Types. An object type $T_{obj} \in \mathbb{T}_{obj}$ describes the properties of a class of objects.[1] It is defined as a set of field names $\{F_1, F_2, \ldots, F_n\}$ each of which is associated with a type $T_i \in \mathbb{T}$, where $1 \leq i \leq n$.

An object type T_{sub} can be a subtype of one or more object types T_{super}, denoted as $T_{sub} \sqsubseteq T_{super}$. The relation $\sqsubseteq$ is transitive, i.e. $T_1 \sqsubseteq T_2 \wedge T_2 \sqsubseteq T_3 \Rightarrow T_1 \sqsubseteq T_3$ and reflexive, i.e. $T_{obj} \sqsubseteq T_{obj}$. Based on these properties, we define

$$T_{obj}^* = \bigcup_{\forall\, T_i \in \mathbb{T}_{obj}\,:\, T_{obj} \sqsubseteq T_i} T_i$$

to be the set of all defined and inherited field names of an object type T_{obj}.[2]

An *object*, $v_{obj} \in \mathbb{V}_{obj}$, is defined as the structure $\langle id, \Omega \rangle$, where id is the object's unique and immutable identifier and $\Omega = \{\mu \mid \mu : T_{obj} \to \mathbb{V}\}$ is a set

---

[1] Due to space limitations, we omit the discussion of methods in this paper.

[2] Note that the precise definition of the set T_{obj}^* depends on the model of inheritance used by the object database. Since our object model does not preclude any inheritance model, different systems may return different sets.

of mappings. Each mapping $\mu : (F_1 = v_1, F_2 = v_2, \ldots, F_n = v_n)$ is a function relating field names $F_i \in T_{obj}$ to values $v_i \in \mathbb{V}$ with the restriction that $\mu(F_i) \vdash T_i$. We say a mapping μ satisfies T_{obj}, denoted by $\mu \models T_{obj}$, iff $\forall F_i \in T_{obj}, \exists v \in \mathbb{V} : \mu(F_i) = v$. An object $v_{obj} = \langle id, \Omega \rangle$ is said to be an instance of T_{obj}, denoted as $v_{obj} \vdash T_{obj}$, iff $\forall T_i \in \mathbb{T}_{obj} : T_{obj} \sqsubset T_i, \exists \mu \in \Omega : \mu \models T_i$. Mappings correspond to the information units introduced earlier.

Both object types and objects can be created, modified and deleted. Due to space limitations, we limit our presentation to the *dress, strip* and *browse* operations that are specific to our object data model. A more comprehensive discussion can be found in [22]. The *dress* and *strip* operations are used respectively to add or remove information units to or from an object, while the *browse* operation computes a mapping that represents the object in the context of the given type.

$$dress(\langle id, \Omega \rangle, T_{obj}) : \text{if } \not\exists\, \mu \in \Omega : \mu \models T_{obj} \text{ then } \Omega := \Omega \cup \{\mu_{new}\} \text{ end}$$
$$strip(\langle id, \Omega \rangle, T_{obj}) : \text{if } \exists\, \mu \in \Omega : \mu \models T_{obj} \text{ then } \Omega := \Omega \backslash \{\mu\} \text{ end}$$
$$browse(\langle id, \Omega \rangle, T_{obj}) : \text{return } \mu : \mu \models T_{obj}^*$$

Structured Types. A structured type $T_{struct} \in \mathbb{T}_{struct}$ describes the structure of literals. Similar to object types, structured types are defined as a set of field names $\{F_1, F_2, \ldots, F_n\}$ where each F_i is associated with a type $T_i \in \mathbb{T}$, where $1 \leq i \leq n$. In contrast to object types, structured types cannot define methods and there is no notion of subtyping or inheritance.

Since a *structured value* or *struct*, $v_{struct} \in \mathbb{V}_{struct}$, has no identity, it is simply defined as a mapping $\mu : T_{struct} \rightarrow \mathbb{V}$, denoted as $(F_1 = v_1, F_2 = v_2, \ldots, F_n = v_n)$, where $\mu(F_i) \vdash T_i$. We say a structured value $v_{struct} = \mu$ is an instance of T_{struct}, denoted as $v_{struct} \vdash T_{struct}$, iff $\mu \models T_{struct}$, where $\mu \models T_{struct} \Leftrightarrow \forall F_i \in T_{struct}, \exists v \in \mathbb{V} : \mu(F_i) = v$.

Extent Types. An extent type $T_{ext} \in \mathbb{T}_{ext}$ describes a collection of values in terms of its bulk behaviour and the type of its members. Accordingly, it is defined as a structure $\langle bulk, T \rangle$, where $bulk \in \{\text{set}, \text{bag}, \text{ranking}, \text{sequence}\}$ and $T \in \mathbb{T}$.

An *extent value* or *extent*, $v_{ext} \in \mathbb{V}_{ext}$, for an extent type $T_{ext} = \langle bulk, T \rangle$ is a finite collection of values, denoted as $v_{ext} = \langle\!\langle v_1, v_2, \ldots, v_n \rangle\!\rangle$. Corresponding to the four bulk behaviours introduced above, we distinguish set, bag, ranking and sequence extent values, depending on whether they are ordered and allow duplicates. We denote a set (unordered, no duplicates) as $v_{set} = \{v_1, v_2, \ldots, v_n\}$, a bag (unordered, duplicates) as $v_{bag} = \rfloor v_1, v_2, \ldots, v_n \lfloor$, a ranking (ordered, no duplicates) as $v_{rnk} = \lceil v_1, v_2, \ldots, v_n \rceil$, and a sequence (ordered, duplicates) as $v_{seq} = [v_1, v_2, \ldots, v_n]$. An extent value v_{ext} is an instance of an extent type $T_{ext} = \langle bulk, T \rangle$, denoted as $v_{ext} \vdash T_{ext}$, iff its behaviour matches $bulk$ and $\forall v \in v_{ext} : v \vdash T$. We will discuss the operations defined over collections of values in the next section.

Example. For the example introduced in Fig. 1, the representation of a database containing objects id_1 and id_2 based on the formal object data model is given by

$$\mathbb{V} = \{\langle id_1, \{\mu_1^{contact}, \mu_1^{person}\}\rangle, \langle id_2, \{\mu_2^{contact}, \mu_2^{organisation}\}\rangle\},$$

where

$$\mu_1^{contact} : (name = \texttt{"Fred Bloggs"}, phones = \{\texttt{"555-123456"}\})$$
$$\mu_1^{person} : (title = \texttt{"Mr"})$$
$$\mu_2^{contact} : (name = \texttt{"ACME"}, phones = \{\texttt{"555-234567"}\})$$
$$\mu_2^{organisation} : (address = (street = \texttt{"Main St."}, number = 9), employees = \{id_1\}).$$

4   Collection Algebra

We now present the algebra associated with our model. Since, for the most part, its operators apply to collections of values, i.e. extent values, we refer to it as a collection algebra. Our algebra is an extension of traditional set algebra as it introduces functionality specific to object data management and provides support for collections other than sets. However, in order to define how these operators manipulate collections of values, we first need to specify their behaviour in terms of the type system of our object data model.

Table 1. Most-specific types

(a) Base types

$\sqcup$	BOOLEAN	INTEGER	REAL	DATE	STRING
BOOLEAN	BOOLEAN	$\perp$	$\perp$	$\perp$	STRING
INTEGER		INTEGER	REAL	$\perp$	STRING
REAL			REAL	$\perp$	STRING
DATE				DATE	STRING
STRING					STRING

(b) Extent types

$\sqcup$	set	bag	ranking	sequence
set	set	set	set	set
bag		bag	set	bag
ranking			ranking	ranking
sequence				sequence

Most-specific Type. We define the most-specific type of two types T_1 and T_2, denoted as $\widehat{T} = T_1 \sqcup T_2$, where $T_1, T_2 \in \mathbb{T}_i$ and $\mathbb{T}_i \in \mathbb{T}^*$. In the case that $\mathbb{T}_i = \mathbb{T}_{base}$, the most-specific type of two base types is defined by Tab. 1(a), where $\perp$ stands for undefined. The most-specific type of two object types $T_1, T_2 \in \mathbb{T}_{obj}$ is defined as

$$\widehat{T} = T_1 \sqcup T_2 \leftrightarrow T_1 \sqsubset \widehat{T} \wedge T_2 \sqsubset \widehat{T} \wedge (\nexists T_i \neq \widehat{T} : T_1 \sqsubset T_i \sqsubset \widehat{T} \wedge T_2 \sqsubset T_i \sqsubset \widehat{T}).$$

If $\mathbb{T}_i = \mathbb{T}_{struct}$, the most-specific type of two structured types T_1 and T_2 is defined as follows. Let $T_1 = \{F_1^1, F_2^1, \ldots, F_n^1\}$ with associated types T_i^1, where $1 \leq i \leq n$ and $T_2 = \{F_1^2, F_2^2, \ldots, F_m^2\}$ with associated types T_j^2, where $1 \leq j \leq m$. If $n = m$ and $\forall_{1 \leq k \leq n} F_k^1, F_k^2 : F_k^1 = F_k^2$, then $\widehat{T} = T_1 \sqcup T_2$ is given as the set of field names $\{F_1 = F_1^1, F_2 = F_2^1, \ldots, F_n = F_n^1\}$ with associated types $T_i = T_i^1 \sqcup T_i^2, 1 \leq i \leq n$. Finally, in the case that $T_1 = \langle bulk_1, T_1'\rangle$ and $T_2 = \langle bulk_2, T_2'\rangle \in \mathbb{T}_{ext}$, the most-specific type of two extent types is given by the structure $\widehat{T} = \langle bulk, T'\rangle$, where $bulk = bulk_1 \sqcup bulk_2$, according to Tab. 1(b) and $T' = T_1' \sqcup T_2'$. In all other cases, the most-specific type of two types is undefined ($\perp$).

Type Compatibility. Two types T_i and T_j are said to be compatible, denoted as $T_i \sim T_j$, iff $T_i \sqcup T_j \neq \bot$.

Support Operations. Finally, we introduce the following operations to support the definition of operators over ordered collections. For an ordered collection $C = \langle\!\langle x | C' \rangle\!\rangle$, the $|$ operator decomposes C into its first element x and the ordered collection of the remaining elements C'. The operation $append(C, x)$: $(\mathsf{coll}[T], T) \to \mathsf{coll}[T]$ inserts an element x at the end of an ordered collection C. The operation $remove(C, x)$: $(\mathsf{coll}[T], T) \to \mathsf{coll}[T]$ removes the element x with the smallest index from the ordered collection C.

Note that we will use the set representation of bags in some of the following definitions, where $\lfloor 1, 1, 1, 2, 2, 3 \rceil \equiv \{(1, 3), (2, 2), (3, 1)\}$. Then we use $x \in_{bag} B$ to denote the membership of x in a bag B and $(x, n) \in_{set} B$ to denote the membership of (x, n) in the set representation of B where n is an integer giving the number of occurrences of x. The full definition of collection membership $\in : (T, \mathsf{coll}[T]) \to \textsc{boolean}$, is given below.

$$x \in_{set} S \;= x \in S \qquad\qquad x \in_{bag} B = \exists n : (x, n) \in_{set} B \wedge n > 0$$
$$x \in_{rnk} R = \exists i : R[i] = x \qquad x \in_{seq} Q = \exists i : Q[i] = x$$

Finally, we also include a definition of bag addition here, which will be used to define other operators over bags that are part of the collection algebra.

$$B_1 \uplus B_2 = \{(x, y) \mid \exists n_1, n_2 : (x, n_1) \in_{set} B_1 \wedge (x, n_2) \in_{set} B_2 \wedge n = n_1 + n_2\}$$

Collection Operations. The extent operation, $\otimes : T \to \mathsf{coll}[T]$, where $T \in \mathbb{T}_{obj}$, returns all objects v_{obj} in the databases, such that $v_{obj} \vdash T$.

The union, $\cup : (\mathsf{coll}[t_1], \mathsf{coll}[t_2]) \to \mathsf{coll}[t_1 \sqcup t_2]$, of two collections is defined as follows.

$$S_1 \cup_{set} S_2 = \{x \mid x \in_{set} S_1 \vee x \in_{set} S_2\}$$
$$B_1 \cup_{bag} B_2 = \{(x, n) \mid \exists n_1, n_2 : (x, n_1) \in_{set} B_1 \wedge (x, n_2) \in_{set} B_2 \wedge n = max(n_1, n_2)\}$$
$$R_1 \cup_{rnk} R_2 = \begin{cases} R_1 & \text{if } R_2 = \emptyset \\ append(R_1, x) \cup_{rnk} R'_2, \text{where } R_2 = \lceil x | R'_2 \rceil & \text{otherwise} \end{cases}$$
$$Q_1 \cup_{seq} Q_2 = \begin{cases} Q_1 & \text{if } Q_2 = \emptyset \\ append(Q_1, x) \cup_{seq} Q'_2, \text{where } Q_2 = [x | Q'_2] & \text{otherwise} \end{cases}$$

The definition of the intersection, $\cap : (\mathsf{coll}[t_1], \mathsf{coll}[t_2]) \to \mathsf{coll}[t_1 \sqcup t_2]$, of two collections is given below.

$$S_1 \cap_{set} S_2 = \{x \mid x \in_{set} S_1 \wedge x \in_{set} S_2\}$$
$$B_1 \cap_{bag} B_2 = \{(x, n) \mid \exists n_1, n_2 : (x, n_1) \in_{bag} B_1 \wedge (x, n_2) \in_{bag} B_2 \wedge n = min(n_1, n_2)\}$$
$$R_1 \cap_{rnk} R_2 = \begin{cases} \emptyset & \text{if } R_1 = \emptyset \\ \lceil x | (R'_1 \cap_{rnk} R_2) \rceil, \text{where } R_1 = \lceil x | R'_1 \rceil & \text{if } x \in_{rnk} R_2 \\ R'_1 \cap_{rnk} R_2, \text{where } R_1 = \lceil x | R'_1 \rceil & \text{otherwise} \end{cases}$$
$$Q_1 \cap_{seq} Q_2 = \begin{cases} \emptyset & \text{if } Q_1 = \emptyset \\ [x | (Q'_1 \cap_{seq} remove(Q_2, x))], \text{where } Q_1 = [x | Q'_1] & \text{if } x \in_{seq} Q_2 \\ Q'_1 \cap_{seq} Q_2, \text{where } Q_1 = [x | Q'_1] & \text{otherwise} \end{cases}$$

The following definition specifies the difference, $- : (\text{coll}[t_1], \text{coll}[t_2]) \rightarrow \text{coll}[t_1]$, of two collections.

$$S_1 -_{set} S_2 = \{x \mid x \in_{set} S_1 \wedge x \notin_{set} S_2\}$$

$$B_1 -_{bag} B_2 = \{(x, n) \mid \exists n_1 : (x, n_1) \in_{set} B_1 \wedge$$
$$((x \notin_{bag} B_2 \wedge n = n_1) \vee \exists n_2 : (x, n_2) \in_{set} B_2 \wedge n = n_1 - n_2)\}$$

$$R_1 -_{rnk} R_2 = \begin{cases} R_1 & \text{if } R_2 = \emptyset \\ remove(R_1, x) -_{rnk} R_2', \text{where } R_2 = \lceil x|R_2' \rceil & \text{if } x \in_{rnk} R_1 \\ R_1 -_{rnk} R_2', \text{where } R_2 = \lceil x|R_2' \rceil & \text{otherwise} \end{cases}$$

$$Q_1 -_{seq} Q_2 = \begin{cases} Q_1 & \text{if } Q_2 = \emptyset \\ remove(Q_1, x) -_{seq} Q_2', \text{where } Q_2 = [x|Q_2'] & \text{if } x \in_{seq} Q_1 \\ Q_1 -_{seq} Q_2', \text{where } Q_2 = [x|Q_2'] & \text{otherwise} \end{cases}$$

Selection. The selection operation, $\sigma : (\text{coll}[t], t \rightarrow \text{BOOLEAN}) \rightarrow \text{coll}[t]$, forms a subcollection of a given collection C that only contains elements that satisfy a predicate p. Using the reduce operation ($\triangleright$), which will be introduced later, it is defined as follows.

$$\sigma_{set} \, p \, S = \{x \mid x \in_{set} S \wedge p(x) = \text{true}\}$$
$$\sigma_{bag} \, p \, B = \{(x, n) \mid (x, n) \in_{set} B \wedge p(x) = \text{true}\}$$
$$\sigma_{rnk} \, p \, R = \triangleright_{rnk} \lambda(x, R').(\text{if } p(x) \text{ then } \lceil x \rceil \cup_{rnk} R' \text{ else } R') \, \emptyset \, R$$
$$\sigma_{seq} \, p \, Q = \triangleright_{seq} \lambda(x, Q').(\text{if } p(x) \text{ then } [x] \cup_{seq} Q' \text{ else } Q') \, \emptyset \, Q$$

Map Operations. Our algebra also supports map operations that apply a given function f to all members of a collection C and return a new collection containing the results of this function application. The general map operator, $\triangleleft : (\text{coll}[t_1], t_1 \rightarrow t_2) \rightarrow \text{coll}[t_2]$, is given as follows.

$$\triangleleft_{set} \, f \, S = \{f(x) \mid x \in_{set} S\}$$
$$\triangleleft_{bag} \, f \, B = \triangleright_{bag} \lambda((x, n), B').(\{(f(x), n)\} \uplus B') \, \emptyset \, B$$
$$\triangleleft_{rnk} \, f \, R = \triangleright_{rnk} \lambda(x, R').(\lceil f(x) \rceil \cup_{rnk} R') \, \emptyset \, R$$
$$\triangleleft_{seq} \, f \, Q = \triangleright_{seq} \lambda(x, Q').([f(x)] \cup_{seq} Q') \, \emptyset \, Q$$

The navigation operation, $\cdot : (\text{coll}[T], F_i) \rightarrow \text{coll}[T_i]$, where $T \in \mathbb{T}_{obj}$, $F_i \in T$ and $\mu(F_i) \vdash T_i$, is a special case of a map operation that substitutes each object $x = \langle id, \Omega \rangle$ with the value of its field F_i, denoted as $x.F_i = \mu(F_i)$, where $\mu \in \Omega$.

$$S \cdot_{set} F = \triangleleft_{set} \lambda x.(x.F) \, S \qquad B \cdot_{bag} F = \triangleleft_{bag} \lambda x.(x.F) \, S$$
$$R \cdot_{rnk} F = \triangleleft_{rnk} \lambda x.(x.F) \, S \qquad Q \cdot_{seq} F = \triangleleft_{seq} \lambda x.(x.F) \, S$$

Reduce Operations. The last group of operators provided in our algebra are reduce operations which, given an aggregation function f and a default value v, compute one or more aggregated values over a collection C. The general reduce operator, $\rhd : (\mathsf{coll}[t_1], ((t_1, t_2) \to t_2), t_2) \to t_2$ is defined as follows.

$$\rhd_{set} \; f \; v \; S = \text{if } S = \emptyset \text{ then } v \text{ else } f(x, \rhd_{set} \; f \; v \; S'), \text{where } S = S' \cup_{set} \{x\}$$
$$\rhd_{bag} \; f \; v \; B = \text{if } B = \emptyset \text{ then } v \text{ else } f(x, \rhd_{bag} \; f \; v \; B'), \text{where } B = B' \uplus \{(x, 1)\}$$
$$\rhd_{rnk} \; f \; v \; R = \text{if } R = \emptyset \text{ then } v \text{ else } f(x, \rhd_{rnk} \; f \; v \; R'), \text{where } R = \lceil x \rceil \cup_{rnk} R'$$
$$\rhd_{seq} \; f \; v \; Q = \text{if } Q = \emptyset \text{ then } v \text{ else } f(x, \rhd_{seq} \; f \; v \; Q'), \text{where } Q = \lceil x \rceil \cup_{rnk} Q'$$

Examples. Based on the example given in Fig. 1, assume we want to find the names of all employees working for the "ACME" company. Then this query could be expressed as follows.

$$(\sigma_{\text{name}=\texttt{"ACME"}}(\otimes\text{ORGANISATION})) \cdot \text{employees} \cdot \text{name}$$

Another example is the following query to retrieve the names of the organisations for which "Fred Bloggs" works. Note that we have split it into two steps purely for the sake of legibility.

$$fred := \sigma_{\text{name}=\texttt{"Fred Bloggs"}}(\otimes\text{PERSON})$$

$$\sigma_{fred \subseteq \text{employees}}(\otimes\text{ORGANISATION}) \cdot \text{name}$$

Apart from the operators presented in this section, our algebra provides further functionality that had to be omitted due to space limitations. A complete overview of our collection algebra can be found in [22].

5   Implementation

Based on the formal definitions given in the previous sections, we have specified an application programming interface (API) and realised a proof-of-concept implementation. The aim of the proposed API is to serve as a standard for uniform access to object databases, rather than as a standard for application development. As a consequence, our API is quite low-level and procedural. Its main concepts are two interface classes that respectively define the methods to manage and query data according to the object data model and algebra. The signatures of the most commonly used methods of the first interface class are outlined in Tab. 2. These methods allow types to be created and instantiated, and their instances to be read, manipulated and deleted.

For example, an object type can be created with the createObjectType method by providing its name and a list of attribute types. Attributes may be of base, structured, object or extent types, which are commonly generalised as Type. An object is created using createObject and dressed with an object type using the dress method which takes the object to be dressed and an object type as argument. Given such an object and its type, attribute values may be read and written with the get/setAttributeValues methods. Finally, an object may be deleted with the deleteObject method. All other types and their instances are managed similarly.

Table 2. Signatures of API methods

createObjectType(Transaction, String, Type[]): ObjectType
createStructuredType(Transaction, String, Type[]): StructuredType
createExtentType(Transaction, String, BulkType, Type): ExtentType
getType(String): Type
createObject(Transaction): Identifier
dressObject(Transaction, Identifier, ObjectType)
stripObject(Transaction, Identifier, ObjectType)
deleteObject(Transaction, Identifier)
getAttributeValues(Transaction, Identifier, ObjectType): Object[]
setAttributeValues(Transaction, Identifier, ObjectType, Object[])
createExtent(Transaction, ExtentType): ExtentValue
deleteExtent(Transaction, ExtentValue)

The interface of the algebra is based on the iterator model [23] and thus follows a language-integrated rather than a declarative approach. The signatures of the algebra operators defined in the previous section are shown in Tab. 3. Additionally, our interface provides a scan method that, given an ExtentValue, returns an iterator. Thus, the scan method interfaces between the collection representation of the object data model and the one used in the algebra. The signatures of the remaining operator methods closely correspond to the formal definitions of Sect. 4 and therefore require no further explanation. Note that all of these methods take one or more iterators as input and return one iterator as output. Therefore, operators can be arbitrarily nested to form complex queries.

As a proof-of-concept, we show how the API was implemented using Berkeley DB Java Edition which is a light-weight key-value database providing direct access to its data structures. Due to the nature of our interface, we wanted to avoid the complexity of interacting with a relational or object database system. While this might sound counter-intuitive, it is motivated by the fact that we do not propose an interface for application development and, therefore, do not believe it should be implemented "on top" of an existing database interface. Rather, vendors should offer the proposed interface as an alternative that supports use-cases such as benchmarking and data exchange.

Table 3. Signatures of algebra operators

scan(ExtentValue): Iterator
map(Iterator, Function): Iterator
reduce(Iterator, Function, Object): Iterator
selection(Iterator, Predicate): Iterator
navigate(Iterator, String): Iterator
union(Iterator, Iterator): Iterator
intersection(Iterator, Iterator): Iterator
difference(Iterator, Iterator): Iterator

In order to store information about the different types, we use separate databases[3] for base, structured, object and extent types. These four databases constitute the metadata over the persistent data and their record layouts are shown in Fig. 2. As every type is identified by a unique name, we map these names to UUID values which are used as database keys. For object types shown in Fig. 2(a), we store a header (grey fields) containing the field and supertype count as well as the offset for the supertypes within the record. We then have a sequence of (Position, ^Type) pairs describing the type's attributes. Position is used for schema evolution, while ^Type is a type reference represented as the UUID of the attribute type. A sequence of UUID values referring to a type's supertypes forms the end of such records. The numbers in parenthesis show the sizes of each record part in bytes. Figure 2(b) shows how base types are described by a Type which encodes the basic type from $\mathbb{T}_{base}$. A record describing a structured type is shown in Fig. 2(c). It consists of a header containing the field count and a sequence of ^Type containing the UUID values of the field types. As shown in Fig. 2(d), extent types are stored as an encoded bulk type such as set, bag, sequence or ranking, and the UUID of the type describing the extent members.

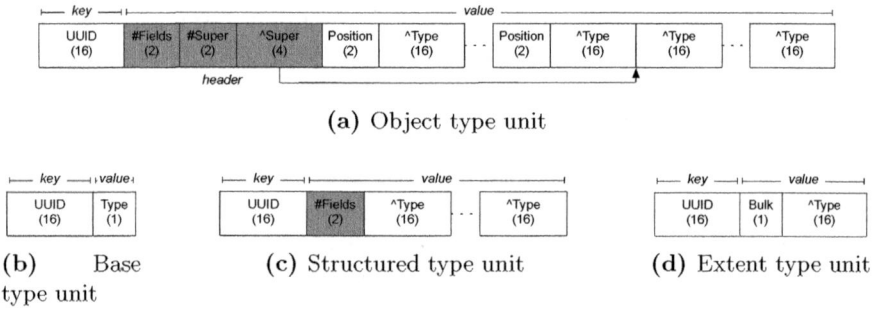

(a) Object type unit

(b)   Base
type unit

(c) Structured type unit

(d) Extent type unit

Fig. 2. Record layouts in the type metadata databases

In addition to the metadata, a user partition contains the objects, their information units and the extent values. For each object type, we create a separate database containing all of its instances. The entries of such databases start with the instance object's identifier encoded as a UUID key, followed by a value part as shown in Fig. 3. The value part contains the information unit's attribute values. Internally, we divide an entry's value part into a fixed-size and a variable-size part. For variable-size attributes such as strings, we store their length and a pointer to the beginning of the variable-size part following the fixed-size part (light grey). This record layout enables the execution of some schema evolution operations without having to re-write all instances of the type under change.

The dress types database shown in Fig. 4 is used to keep track of all information units that belong to an object. In this database, we map the object's UUID

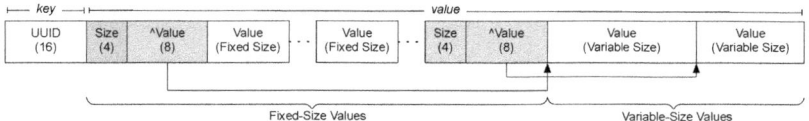

Fig. 3. Record layout for object information units

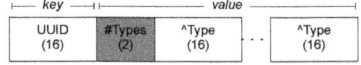

Fig. 4. Record layout for object dress types

to a sequence of UUID values referring to all types an object has been dressed with. This database duplicates information that could be found by accessing all type extents, however, we use it as an index to accelerate look-up operations.

Each extent is stored in its own database. Depending on the bulk type, additional index structures such as Berkeley DB's secondary databases are employed for fast access to extent members. The members are UUID values in the case of objects and extents or the actual values in the case of base and structured values.

6   Discussion

We now discuss and position our work with respect to the related approaches that were introduced in Sect. 2. The object data model that we presented in this paper can be classified as an evolution of the ODMG 3.0 data model. The ODMG modelling primitives of objects and literals correspond to object and structured types in our model. The distinction of whether information is modelled as an identifiable object or an inlined value is present in most object databases. For example, the Versant Object Database (VOD) introduced the notion of first-class and second-class objects, while Objectivity/DB uses the concept of embedded objects to support this feature. As a consequence, we believe that any new object database standard should also include these capabilities. Finally, we note that the collection types defined in our model are slightly different from the ones offered by the ODMG model. Nevertheless, we share the conviction that different collection types and their associated operations are an essential part of an object data model.

The approach that is currently proposed as the next-generation object databases standard takes an altogether different stance in this respect. Instead of defining the characteristics of a standard object data model, their data model decomposes objects into triples that are used to represent all information. While this model is very flexible and easily formalised, it is too general and lacks specificity for the domain of object databases. Our model acknowledges the importance of a formal specification as the foundation of consistent semantics, however we position it differently in terms of the trade-off between flexibility and specificity. Based on object-slicing, our approach supports different models of inheritance and instantiation. At the same time, its type model and collection algebra are truly object-oriented.

When defining a standard, there are different objectives that can be taken into consideration. For example, the ODMG 3.0 standard has been defined to provide better support for unified application development and portability. The goal of the interface proposed in this paper is different as it was designed to facilitate standardised evaluation of object databases in terms of benchmarking or as a format for data exchange. Consequently, our application programming interface does not provide transparent persistence that is nowadays the standard for object database application development. Nevertheless, we believe that the adoption of our proposal is likely as many vendors already offer lower-level interfaces to their databases, e.g. Versant's JVI Fundamental Binding [24].

7   Conclusion

We have presented an object data model that uses object-slicing to support different styles of inheritance and instantiation. We have defined the model formally and used this specification as the basis for a collection algebra that provides query facilities in the context of our object data model. Finally, we have proposed an interface that supports both uniform access and querying of object data that is represented according to the proposed model. As a proof-of-concept, the interface that is intended for benchmarking and data exchange has been implemented using Berkeley DB Java Edition.

As future work, we plan to experiment with different object-slicing strategies. In this paper, we have assumed a one-to-one correspondence between object classes and information units. However, if an object database does not provide multiple inheritance nor multiple instantiation, this assumption might be unreasonable and lead to increased complexity. To experiment with this, we plan to implement our interface based on different existing object databases. At the same time, this will help to demonstrate its value for benchmarking and data exchange.

References

1. Atkinson, M.P., Bancilhon, F., DeWitt, D.J., Dittrich, K.R., Maier, D., Zdonik, S.B.: The Object-Oriented Database System Manifesto. In: Building an Object-Oriented Database System: The Story of O_2, pp. 3–20. Morgan Kaufmann, San Francisco (1992)
2. Dearle, A., Kirby, G.N.C., Morrison, R.: Orthogonal Persistence Revisited. In: Proc. Intl. Conf. on Object Databases (ICOODB), pp. 1–23 (2009)
3. Greene, R.: OODBMS Architectures: An Examination of Implementations. Technical report, Versant Corp. (2006)
4. Cattell, R.G.G., Skeen, J.: Object Operations Benchmark. ACM Trans. Database Syst. 17(1), 1–31 (1992)

5. Carey, M.J., DeWitt, D.J., Naughton, J.F.: The OO7 Benchmark. In: Proc. Intl. Conf. on Management of Data (SIGMOD), pp. 12–21 (1993)
6. Cattell, R.G.G., Barry, D.K., Berler, M., Eastman, J., Jordan, D., Russell, C., Schadow, O., Stanienda, T., Velez, F. (eds.): The Object Data Standard: ODMG 3.0. Morgan Kaufmann, San Francisco (2000)
7. Card, M.: Next-Generation Object Database Standardization. Technical report, Object Management Group (OMG) (2007)
8. Adamus, R., Habela, P., Kaczmarski, K., Lentner, M., Stencel, K., Subieta, K.: Stack-Based Architecture and Stack-Based Query Language. In: Proc. Intl. Conf. on Object Databases (ICOODB), pp. 77–95 (2008)
9. Frost, R.A.: Binary-Relational Storage Structures. Comput. J. 25(3), 358–367 (1982)
10. Papakonstantinou, Y., Garcia-Molina, H., Widom, J.: Object Exchange Across Heterogeneous Information Sources. In: Proc. Intl. Conf. on Data Engineering (ICDE), pp. 251–260 (1995)
11. Versant Corp.: Versant Object Database Fundamentals Manual, Release 8.0 (2009)
12. Box, D., Hejlsberg, A.: The LINQ Project. Technical report, Microsoft Corporation (2005)
13. Meijer, E., Beckman, B., Bierman, G.: LINQ: Reconciling Object, Relations and XML in the.NET Framework. In: Proc. Intl. Conf. on Management of Data (SIG-MOD), pp. 706–706 (2006)
14. Paterson, J., Edlich, S., Hörning, H., Hörning, R.: The Definitive Guide to db4o. Apress (2006)
15. Martin, J., Odell, J.J.: Object-Oriented Analysis and Design. Prentice-Hall, Inc., Englewood Cliffs (1992)
16. Parsons, J., Wand, Y.: Emancipating Instances from the Tyranny of Classes in Information Modeling. ACM Trans. Database Syst. 25(2), 228–268 (2000)
17. Ra, Y.G., Kuno, H.A., Rundensteiner, E.A.: A Flexible Object-Oriented Database Model and Implementation for Capacity-Augmenting Views. Technical Report CSE-TR-215-94, University of Michigan (1994)
18. Kuno, H.A., Ra, Y.G., Rudensteiner, E.A.: The Object-Slicing Technique: A Flexible Object Representation and its Evaluation. Technical Report CSE-TR-241-95, University of Michigan (1995)
19. Fishman, D.H., Beech, D., Cate, H.P., Chow, E.C., Connors, T., Davis, J.W., Derrett, N., Hoch, C.G., Kent, W., Lyngbæk, P., Mahbod, B., Neimat, M.A., Ryan, T.A., Shan, M.C.: Iris: An Object-Oriented Database Management System. ACM Trans. Office Info. Syst. 5(1), 48–69 (1987)
20. Bernstein, P.A., Halevy, A.Y., Pottinger, R.A.: A Vision for Management of Complex Models. SIGMOD Rec. 29(4), 55–63 (2000)
21. Bauer, C., King, G.: Java Persistence with Hibernate. Manning Publications Co. (2006)
22. Würgler, A.P.: OMS Development Framework: Rapid Prototyping for Object-Oriented Databases. PhD thesis, ETH Zurich (2000)
23. Graefe, G.: Volcano–An Extensible and Parallel Query Evaluation System. IEEE Trans. on Knowl. and Data Eng. 6(1), 120–135 (1994)
24. Versant Corp.: Java Versant Interface Usage Manual, Release 8.0 (2009)

Data Model Driven Implementation of Web Cooperation Systems with Tricia

Thomas Büchner, Florian Matthes, and Christian Neubert

Technische Universität München, Institute for Informatics,
Boltzmannstr. 3, 85748 Garching, Germany
{buechner,matthes,neubert}@in.tum.de
http://wwwmatthes.in.tum.de

Abstract. We present the data modeling concepts of Tricia, an open-source Java platform used to implement enterprise web information systems as well as social software solutions including wikis, blogs, file shares and social networks. Tricia follows a data model driven approach to system implementation where substantial parts of the application semantics are captured by domain-specific models (data model, access control model and interaction model). In this paper we give an overview of the Tricia architecture and development process and present the concepts of its data model: plugins, entities, properties, roles, mixins, validators and change listeners are motivated and described using UML class diagrams and concrete examples from Tricia projects. We highlight the benefits of this data modeling framework for application developers (expressiveness, modularity, reuse, separation of concerns) and show its impact on user-related services (content authoring, integrity checking, link management, queries and search, access control, tagging, versioning, schema evolution and multilingualism). This provides the basis for a comparison with other model based approaches to web information systems.

Keywords: Data modeling, web framework, web application, software engineering, software architecture, domain specific language.

1 Motivation and Introduction

Developers of enterprise web information systems and social software solutions are faced with a complex technology stack of programming languages (Java, PHP, Python, ...), persistence managers (Hibernate, JPA, JCR, ...), authorization and access control frameworks, template engines (Servlets, JSP, ...) and web form validation solutions. In order to support access via web APIs from third-party applications or stateful, rich, mobile clients (iPhone, Android, ...) even more technologies have to be employed.

These software development approaches suffer from the fact that small changes in the customer requirements (e.g., adding an attribute to a persistent entity, changing the cardinality of an association or changing the access policy for a certain user group) lead to numerous changes in all the layers of the server and

A. Dearle and R.V. Zicari (Eds.): ICOODB 2010, LNCS 6348, pp. 70–84, 2010.

even on the client (e.g., JavaScript code for AJAX validation). These changes are very error-prone because of the mismatches between the type systems and data models involved (object-oriented, relational, tree-based). Based on our industrial experience of the last ten years implementing content management solutions, knowledge management solutions and community platforms we have developed during the last three years a data model driven approach which is supported by Tricia, an open source Java platform [3, 13].

The core of Tricia is an innovative modeling language tailored specifically to the needs of this problem domain. The main idea is to derive all necessary boiler-plate implementation details from model-representations (data, access control, interaction) available at runtime as Java classes and objects. The application developer can thus focus on the pure application-specific business logic which is implemented in Java with language (typing, binding, scoping) and IDE support (auto-completion, refactoring, dependency checking, . . .).

The purpose of this paper is to give an overview of the Tricia software archi-tecture and development process (Section 2) and to present and motivate the concepts of its domain-specific data model (Section 4). In Section 3 we use the running example of a small Wiki application that allows end users to create wikis with wiki pages, comments, tags, etc. Throughout the text we highlight the benefits of this particular data model for application developers and for end users. This is the basis for a comparison with related work on model driven im-plementation of web applications in Section 6. We present exemplary views on the data model generated by model introspection (Section 5). The paper ends with concluding remarks and points to subsequent publications which will de-scribe Tricia's access control model and interaction model that builds on the data model described in this paper.

2   Overview of the Tricia Software Architecture and Development Process

Figure 1 provides an architectural overview of a typical web application imple-mented on the Tricia platform using a notation similar to an UML deployment diagram. Such an application provides HTTP(S) access for its web clients (pos-sibly including AJAX-style asynchronous interactions), a REST-ful web API to allow third parties to query and update the content managed by the application and a *Model Introspection* interface to allow third parties to discover and query the data model, access control model and interaction model implemented by the application. Our current implementation only supports a single server (up to fifteen page requests per second on stock hardware) but the architecture is designed for a scale-out to multiple servers using a cluster database.

A Tricia application requires a Java 1.6 runtime environment on Windows or Linux, a database server, and a Lucene full-text search engine. Currently Tri-cia supports MySQL, Oracle, and for testing purposes the in-memory database HSQLDB. There also exists a prototypical implementation which persists data using the NoSQL database MongoDB.

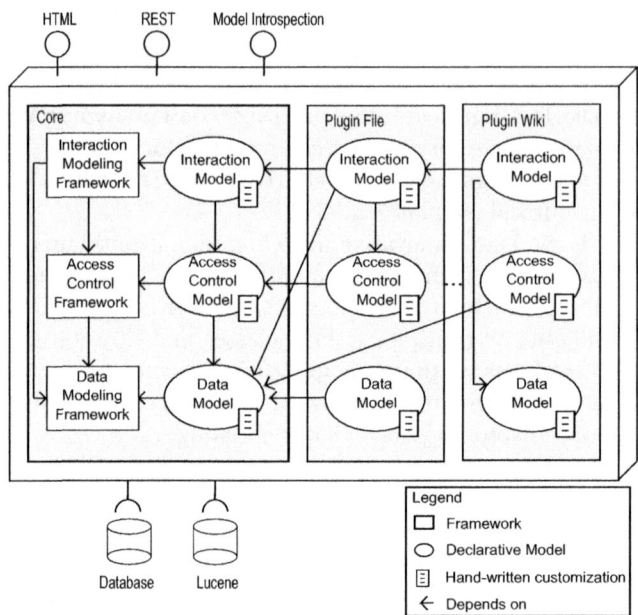

Fig. 1. Architectural overview of a typical web application implemented on the Tricia platform

A Tricia application consists of a *core* and one or more *plugins* that define the application in a modular fashion (see shaded areas in Figure 1). Each plugin specifies the plugins it depends on. Cyclic plugin dependencies are not allowed and are detected at construction time. For example, the plugin Wiki depends on the plugin File, since files may be attached to wiki pages and wiki management thus requires file management.

The core defines abstractions required by virtually all applications of the domain, for example user profiles, groups, memberships, login and registration procedures. The plugins Wiki and File both use such user profiles to identify the last editor of a wiki or a file. Other abstractions provided by the core are discussed in Section 4.4.

Each plugin and the core define a data model, an access control model and an interaction model. Each model defines a fragment of the data structures and behavior of the entire application. These models are expressed by graphical and textual notations (see Section 4) and are available at runtime for introspection (ovals in Figure 1). If necessary, they can be augmented by customizations written in Java (e.g., to express business logic). Models from a plugin P may reference models in P and in plugins imported by P (depicted by arrows in Figure 1). The following rules apply: Interaction models may reference other interaction models, access control models or data models. Access control models may reference other access control models or data models. Data models may reference other data models only. The core is provided as part of the Tricia platform and consists of three layered Java frameworks (c.f. layered architecture in [8]) for data modeling, access control and user interaction (views and controllers). Each framework provides abstractions and extension points, which have to be instantiated or customized in order to build a Tricia application. Frameworks are developed and maintained by the Tricia core developers as part of the core development process, customizations are developed by application developers as part of the application development process [12]. There are two different kinds of customizations. The majority of customizations can be done in a declarative, model driven way. This results in models to be created. For some aspects to be customized it is more convenient to specify them using the full expressive power of the base language, which is Java in our case. An example for this kind of customization is complex business logic. Figure 1 emphasizes the central role of the data modeling framework as the foundation for model driven web application development. Due to space limitations we focus in this paper on the concepts of the data modeling framework. We plan to describe the other frameworks and their meta models in subsequent papers. The following examples should suffice to highlight the use of the application-specific data model in all frameworks:

- For each entity type, the Tricia interaction framework can generate multi-lingual element-oriented CRUD views (create, read, update, delete) and set-oriented table controls. These views may include rich text attributes and media attachments (images and files).
- Associations between entities can be navigated in an element-oriented (via hyperlinks) or set-oriented (declarative queries) style. End users can interactively create full-text and structured queries for entities of a given type or any type (Google-like searches).
- Tricia can also expose these views and controllers as REST-ful web APIs to allow external systems to interact with Tricia applications.
- The Tricia access control framework allows application developers or end users to associate access control policies with entity types or even individual entities. These policies can restrict read, write and administration rights to user groups or to individual users (role based access control or discretionary access control). The policies are enforced automatically at the user interface and at the web API level.
- The Tricia data modeling framework automates the data migration steps necessary after (series of) typical incremental schema changes.

3   A Small Sample Application

In the following, we will introduce the concepts of the Tricia data model step by step using a simple sample application, which allows registered users to manage a collection of wikis. Each of these wikis contains multiple wiki pages. One of the pages of a wiki can be specified as the wiki home page. Wikis and wiki pages are identified by a unique name and a readable, structured and persistent URL.

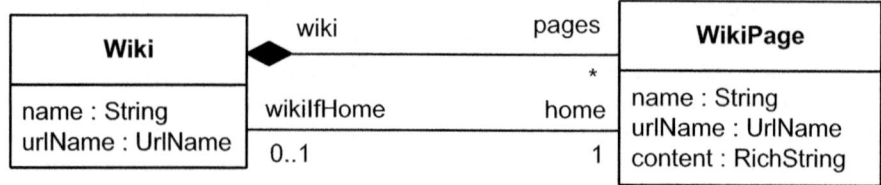

Fig. 2. Overview of an exemplary Tricia data model in a UML-based notation

The content of a wiki page is a rich text (with markup, embedded hyperlinks and attached media files). This Tricia application consists of a plugin with a data model that defines the entities `Wiki` and `WikiPage`.

Tricia data models can be visualized at construction time and at run-time in two different notations:

- A graphical overview notation similar to UML class diagrams (see Figure 2).
- A domain-specific textual syntax which contains all model details (see Figure 3).

The first notation should be self-explanatory and the reader should already get a first idea of the concepts of the DSL used in Figure 3 which are described in the next section.

4 The Data Modeling Concepts of Tricia

As explained in the architecture overview section, the data modeling framework is responsible for the management of persistent and volatile data as specified by the data models of the core and the plugins. Technically speaking, the framework provides (possibly abstract and polymorphic) Java classes for each data modeling concept of Tricia. These classes are instantiated and customized based on the data model of the specific application.

Figure 4 provides an overview of all concepts of the Tricia data modeling framework. In the following subsections we present each of these classes and some of their extension points and illustrate their use with the wiki sample application introduced in the previous section.

4.1 Entities, Properties and Roles

Entities. Tricia domain objects are represented as objects of type `Entity`. The example defines `Wiki` and `WikiPage` entities. Entities have a name, identifying the concept in the data model, and an internationalized label, which is used to generate views for end users. In our example, a `WikiPage` has an internationalized

```
entity Wiki                              entity WikiPage
 label = (en : "Wiki")                    label = (en : "Wiki Page",de : "Wikiseite")
 mandatoryMixins                          mandatoryMixins
  Linkable                                 Commentable
  Seachable                                Linkable
 features                                  Taggable
  name : StringProperty                    Seachable
   maxLength = 255                         features
   isIndexed = false                        name : StringProperty
   isPersistent = true                       maxLength = 255
   label = (en : "Name")                     isIndexed = false
   validate                                  isPersistent = true
    MinimalLengthValidator(length = 1)       label = (en : "Name")
  urlName : UrlNameProperty                  validate
   maxLength = 255                            MinimalLengthValidator(length = 1)
   isIndexed = false                         onChange
   isPersistent = true                        updateUrlName (
   label = (en : "Name in URL")                WikiPage.name,
  pages : ManyRole (WikiPage)                  WikiPage.urlName
   oppositeRole = wiki : OneRole               )
   isCascadeDelete = true                    urlName : UrlNameProperty
   isPersistent = true                        maxLength = 255
  home : OneRole (WikiPage)                    isIndexed = false
   oppositeRole = wikiIfHome : OneRole        isPersistent = true
   isCascadeDelete = false                    label = (en : "Name in URL")
   isOwner = true                            content : RichStringProperty
   isPersistent = true                        maxLength = 16777216
   label = (en : "Home Page")                 isIndexed = false
                                              isPersistent = true
                                              label = (en : "Content")
                                            wiki : OneRole (Wiki)
                                             oppositeRole = pages : ManyRole
                                             isCascadeDelete = false
                                             isOwner = false
                                             isPersistent = true
                                             label = (en : "Wiki")
                                             validate
                                              NotNullOneValidator
                                            wikiIfHome : OneRole (Wiki)
                                             oppositeRole = home : OneRole
                                             isCascadeDelete = false
                                             isOwner = false
                                             isPersistent = true
```

Fig. 3. Detailed Tricia data model in a textual DSL

label with the English text "Wiki Page" as well as the German text "Wiki-seite". The textual representation of the label is as follows (see also Figure 3)

```
entity WikiPage
   label = (en : "Wiki Page",de : "Wikiseite")
```

Properties. Properties of domain objects are represented as objects of type Property. The data modeling framework provides the following predefined basic property types: BooleanProperty, IntProperty, StringProperty, DomainValueProperty, DateProperty, TimestampProperty. Each property

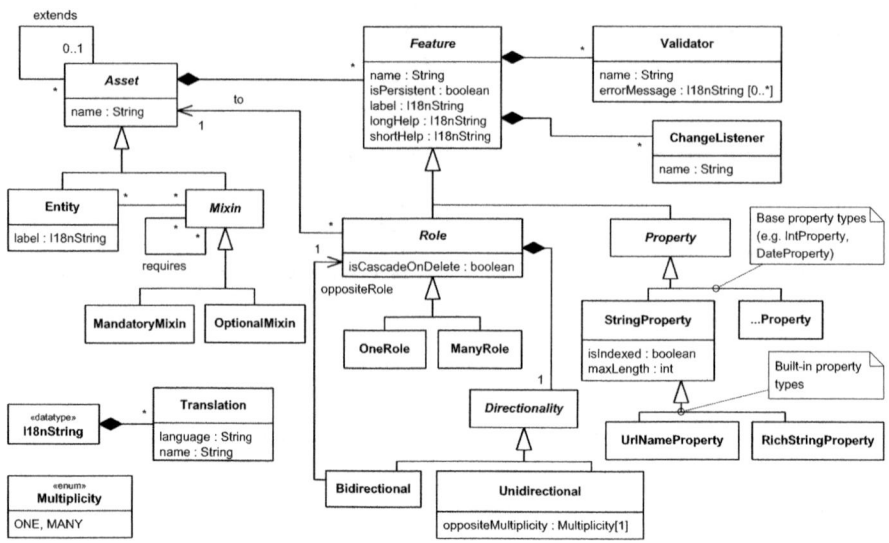

Fig. 4. Concepts of the Tricia data modeling framework

type may introduce certain attributes, which can be customized. For instance, StringProperty represents a character sequence, with a size limited by the maxLength attribute. The attribute isIndexed indicates whether an index shoud be created to speed up value-based queries for that property.

Building on the basic property types (e.g., StringProperty) the Tricia core provides the following domain-specific property types:

- A RichStringProperty is a sub type of StringProperty, which holds HTML content. The implementation of RichStringProperty ensures, that the content does not contain malicious scripts, automatically detects dangling hyperlinks and supports a consistent application-wide URL renaming.
- An UrlNameProperty is used to provide meaningful URLs for domain objects. URLs should match as closely as possible the name of the object, but may be subject to additional constraints due to character set limitations for URLs.
- A PasswordProperty holds encrypted passwords and makes sure that the content of the property is never displayed in views.
- An IdProperty is a sub type of StringProperty with the special semantics of being a unique identifier for an entity. Each entity has a property of type IdProperty.

In our example of Section 3, properties of a *Wiki* are *name* of type String Property and *urlName* of type UrlNameProperty. A *WikiPage* also has the properties *name*, *urlName*, and additionally a *content* property of type RichStringProperty.

Roles. Associations between domain objects are represented in Tricia by modeling the association ends as objects of type `Role`. A role specifies the type of the associated entity, which is represented in the data model framework of Figure 4 by the *to* reference.

There are two kinds of multiplicities: A single-valued association is modeled using the class `OneRole` and a multi-valued association through the class `ManyRole`. The directionality of a role is expressed by the mandatory concept `Directionality`. Bidirectional roles reference the corresponding opposite role. In this case the multiplicity of the counterpart is given implicitly through the type of the opposite role instance (`OneRole` or `ManyRole`). The bidirectional `pages` role from the example data model of Figure 2 is textually represented as follows:

```
pages : ManyRole (WikiPage)
   oppositeRole = wiki : OneRole
```

Since an unidirectional role does not specify an opposite role, its multiplicity cannot be derived and has to be defined explicitly through the `otherMultiplicity` attribute.

The attribute `isCascadeOnDelete` indicates to delete the referenced entities if the owning entity is deleted (c.f. UML composition). In our example, a wiki is used as a container for a set of wiki pages:

```
pages : ManyRole (WikiPage)
   isCascadeDelete = true
```

Features. Properties and roles share some common attributes, which are captured by the abstract super concept `Feature`. Each feature has a name, an internationalized label (c.f. entity attributes), as well as the two internationalized attributes `longHelp` and `shortHelp`. These labels are used in generated views to describe the meaning of a feature to end users in their own language.

The flag `isPersistent` indicates whether the value of a feature is to be stored persistently in a database, by default this flag is set. A non-persistent property can be used for derived values, which are calculated depending on the values of other persistent properties, and can be shown in certain views. The Tricia data modeling framework also supports inheritance, i.e., a derived entity inherits all features of its parent entity. By default, a single-table strategy [11] is used to map the inheritance tree to a single database table.

4.2   Validators

An important aspect of data modeling is the specification of constraints to ensure data integrity. In the Tricia data modeling framework constraints can be modeled through `Validators`. As part of the declarative model, a validator has a name and provides error messages, which are shown to end users in case of a validation failure. The actual algorithm, which computes the state of a validator, is provided as a hand-written customization as introduced in section 2. Validators can be specified for all features, i.e., for roles and properties equally.

As an example, a validator verifies whether the value of a `StringProperty` satisfies a specific pattern (e.g., e-mail address). An example for role validation is to constrain the cardinality of an association. In our example, a wiki page has to be part of a wiki. This can be realized by a role validator applied on role *wiki*:

```
wiki : OneRole (Wiki)
  validate
   NotNullOneValidator
```

The Tricia data modeling framework provides a set of built-in property validators, e.g., `EmailValidator`, `MinimalLengthValidator`, as well as predefined role validators, e.g., `NotNullOneValidator`, `NotEmptyManyValidator`. Validators can be parameterized with values:

```
name : StringProperty
  validate
   MinimalLengthValidator(length = 1)
```

4.3   Change Listener

In the Tricia data modeling framework `ChangeListeners` are used to propagate data model changes through the system. A change listener has a name and is registered on a feature in order to be notified when the value of the feature changes. Change listeners apply for both kinds of features, i.e., roles and properties equally.

For example, a change listener `updateUrlName` can be defined for the name property (StringProperty) of a WikiPage. If the name property is set for a newly created page and no URL is given by the end user, the value of the `name` property is used as default for the URL. In this case, the URL cannot be empty, this is ensured by the validation rule of the name property (cf. `MinimalLengthValidator` in section 4.2).

4.4   Entities and Mixins

The only way of realizing reuse at the data model level introduced so far is the mechanism of inheritance. Since models in Tricia are realized by subclassing framework classes, this mechanism is constrained by having a single inheritance chain, which means that an entity can have only one entity it inherits from. This imposes a severe limitation, and is not sufficient for real-world modeling problems. To enable reuse on a more fine-grained level, Tricia utilizes the concept of *mixins*[1].

Mixins extend entities with additional properties and roles. As shown in Figure 4, the `Entity` and `Mixin` classes are subtypes of the abstract class `Asset`, which provides the capability of having features as introduced in 4.1. Mixins can be assigned to other entities and vice versa, which is expressed by a many-to-many association between `Entity` and `Mixin` as shown in the class diagram in Figure 4.

We distinguish two kinds of mixins, which are realized by the framework classes `MandatoryMixin` and `OptionalMixin`. Mandatory mixins are assigned statically to a certain entity and cannot be removed at runtime. In Table 1 an extract of existing mandatory mixin types and their use by entity classes is shown. These mixins enable fine-grained re-use.

A mixin can depend on other mixins, e.g., a searchable entity (i.e., an entity the mandatory mixin `Searchable` is assigned to) requires to be linkable (have a URL) too, otherwise the asset cannot be accessed if it is shown in a search result list. In this example, it is not permitted to define searchable entities, which are not linkable.

Table 1. Mandatory mixins and their usage in the core and in the Wiki plugin

	Linkable	Searchable	Taggable	Commentable	Versionable
Group	x	x	x		
Membership					
Person	x	x	x		
Principal	x	x			
Comment	x	x			
Search	x	x	x		
Version	x				
Wiki	x	x			
WikiPage	x	x	x	x	x

In contrast, optional mixins can be assigned to objects and can be removed at runtime by end users. An example of an optional mixin is the class `CalendarItem`, which can be assigned to wiki pages. It marks the assigned wiki page as representing a temporal event, which is characterized by additional features such as `startDate`, `endDate`, and `eventCategory`. As opposed to mandatory mixins, this capability is not required for all wiki pages, but can be assigned by end users at runtime. The existence of this mixin type then indicates whether a specific wiki page is displayed in a calendar view, or not.

As shown in Table 1, the Tricia core includes predefined entity types which are essential for the domain of enterprise web applications. They comprise entities for modeling users and user groups: `Person`, `Group`, `Membership`, and `Principal`. These entities are the foundation for the access control framework (see Section 2). Other built-in entites are `Link`, `Comment`, `Version`, which are associated with the respective mandatory mixin types. For example, the mixin `Commentable` establishes a one-to-many association to entities of type `Comment`:

```
mandatoryMixin Commentable          entity Comment
 requires                            label =  (en : "Comment")
  Linkable                           mandatoryMixins
 features                             Linkable
  showComments : BooleanProperty      Searchable
   isIndexed = false                 features
   isPersistent = true                authorName : StringProperty
   label =  (en : "Show Comments")    maxLength = 255
  comments : ManyRole (Comment)       isIndexed = false
   isCascadeDelete = true             isPersistent = true
   isPersistent = true               content : StringProperty
   oppositeRole = commentable : OneRole  maxLength = 16777216
                                       isIndexed = false
                                       isPersistent = true
                                       label =  (en : "Content")
                                       validate
                                        MinimalLengthValidator(length = 5)
                                      creationDate : TimestampProperty
                                       isPersistent = true
                                      commentable : OneRole (Commentable)
                                       isCascadeDelete = false
                                       isOwner = false
                                       isPersistent = true
                                       oppositeRole = comments : ManyRole
```

Fig. 5. Comment and Commentable - textual representation

5   Introspective Implementation

As presented in sections 2 and 4, data models are represented in Tricia as Java classes, which instantiate and customize classes of the data modeling framework. In order to enable a model driven development process, declarative models can be extracted from the Java code by *introspection* [4–6]. Technically speaking, the data modeling framework is an introspective whitebox framework, since it provides annotations in the framework classes, which mark the extension points. Customizations have to follow an introspective programming model, which enables the extraction of the model information. For more details see [4–6].

As already mentioned, Tricia provides different views to visualize the data models. The most generic view presents a data model in a tree structure, which is shown in Figure 6. As it is shown in this Figure, the model view is integrated with the Java source it is derived from.

In order to get an overview of a data model, a graphical presentation similar to UML class diagram notation is provided for Tricia application developers. A screenshot showing the wiki data model is depicted in Figure 7. More details on all features in the graphical view are accessible via the textual representation already introduced as shown in the screenshot.

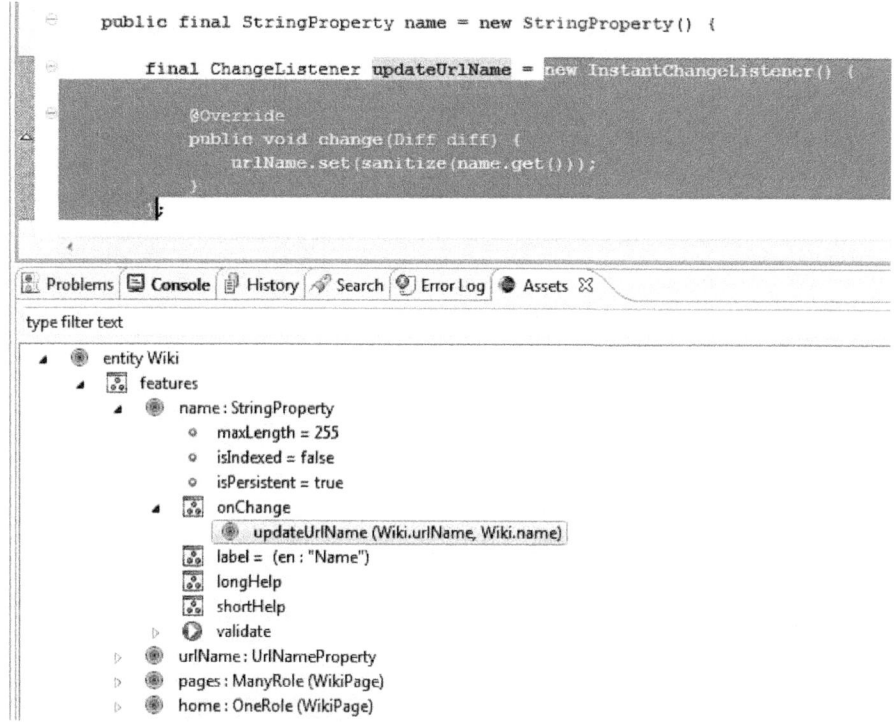

Fig. 6. Tree view of the entity type `Wiki` generated through introspection

6   Related Work

There exist numerous approaches to model driven web development [9, 15–17, 19]. Since the main focus of this paper is on the data model, we will characterize the data modeling capabilities of the following approaches:

- WebML [9] uses a notation which is compatible with classical E/R models and with UML class diagrams. To cope with the requirement of expressing redundant and calculated information, the structural model also offers a simplified, OQL-like query language, by which it is possible to specify derived information.
- UWE [15] uses the graphical UML class diagram notation for data modeling. The main modeling elements used in the conceptual model are: *class* and *association*. Additional features which can be used to semantically improve data models are: *association* and *role* names, *multiplicities*, different forms of associations supported by the UML like *aggregation, inheritance, composition* and *association class*.
- Mod4j [17], WebDSL [19], and MontiWeb [16] specify models using a textual representation, which is transformed by a generator into JPA code.

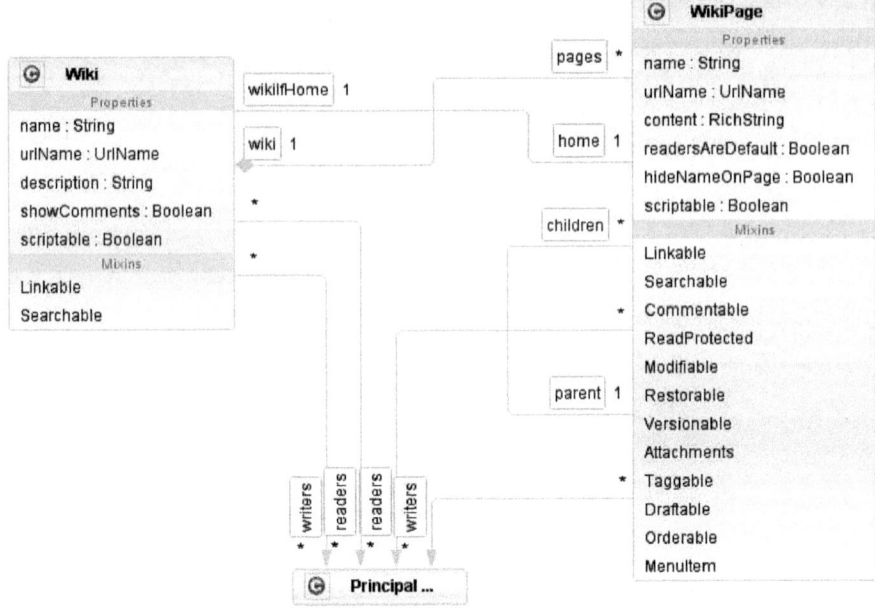

Fig. 7. Introspective Graphical View

None of the existing approaches supports mixin types, which enable reuse as shown in section 4.4.

These existing solutions all use the *generative* approach to model driven development, which means that source code is *generated* from models. What differentiates our approach from these approaches is the idea to extract models from the source code through introspection, which improves the integration of the models with the underlying system [6].

Our introspective approach is closely related to the one introduced in [2] in the sense that declarative model views are extracted from Java source code. In [2] this idea is being applied to behavioral models.

7   Summary

We presented Tricia, an open source Java-based platform for the development of dynamic data intensive enterprise web applications and social software solutions.

We introduced Tricias architecture, its constituents and interfaces. Tricias plugins enable componentized large applications and provide with mixins the basis for supporting software product lines [18] at the data modeling level. Tricia follows a data model driven approach to system implementation. We gave an example of declarative application development based on the data modeling framework in the domain of social software. Tricia provides compile-time and runtime introspection with a strongly typed generic meta-model. We illustrated how textual and graphical views of introspective models facilitate the understanding of complex web applications.

Based on the proposed architecture we built an Enterprise 2.0 tool, which we compared in [7] to existing commercial and open source tools. Our tool consists of 15 plugins with 40 entities and about 500 features and is used in production in several places (e.g., [10, 14]). Our experiences in building and maintaining a system of this size show that a data model driven approach improves the understandability and quality of the system.

Due to space limitations this paper focuses on the data modeling framework. The access control and interaction modeling framework will be subject of subsequent papers.

References

1. Ancona, D., Lagorio, G., Zucca, E.: Jam - designing a Java extension with mixins. ACM Trans. Program. Lang. Syst. 25(5), 641–712 (2003)
2. Balz, M., Striewe, M., Goedicke, M.: Embedding Behavioral Models into Object-Oriented Source Code. In: Liggesmeyer, P., Engels, G., Münch, J., Dörr, J., Riegel, N. (eds.) Software Engineering. LNI, vol. 143, pp. 51–62. GI (2009)
3. Bitbucket - Tricia. Website, http://bitbucket.org/infoasset/tricia-core, (visited on May 30, 2010)
4. Büchner, T.: Introspektive Modellgetriebene Softwareentwicklung. PhD thesis, Technische Universität München (2007)
5. Büchner, T., Matthes, F.: Introspective Model-Driven Development. In: Gruhn, V., Oquendo, F. (eds.) EWSA 2006. LNCS, vol. 4344, pp. 33–49. Springer, Heidelberg (2006)
6. Büchner, T., Matthes, F.: Using Framework Introspection for a Deep Integration of Domain-Specific Models in Java Applications. In: Proceedings of the 1. Workshop des GI-Arbeitskreises Langlebige Softwaresysteme (L2S2): Design for Future - Langlebige Softwaresysteme, pp. 123–135 (2009)
7. Büchner, T., Matthes, F., Neubert, C.: A concept and service based analysis of commercial and open source enterprise 2.0 tools. In: Liu, K. (ed.) KMIS, pp. 37–45. INSTICC Press (2009)
8. Buschmann, G., Meunier, R., Rohnert, H., Sommerlad, P., Stal, M.: Pattern-Oriented Software Architecture: a system of patterns, vol. 1. John Wiley and Sons, Chichester (1996)
9. Ceri, S., Fraternali, P., Bongio, A.: Web Modeling Language (WebML): a modeling language for designing Web sites. Computer Networks 33(1-6), 137–157 (2000)
10. ECHORD (European Clearing House for Open Robotics Development), http://www.echord.info (visited on May 30, 2010)
11. Fowler, M.: Patterns of Enterprise Application Architecture. Addison-Wesley, Reading (2003); With contributions from Rice, D., Foemmel, M., Hieatt, E., Mee, R. and Stafford, R.
12. Froehlich, G., Hoover, H., Liu, L., Sorenson, P.: Designing object-oriented frameworks. University of Alberta, Canada (1998)
13. infoAsset, http://www.infoasset.de, (visited on May 30, 2010)
14. Intranet, Faculty of Informatics, Technical University Munich, http://intranet.in.tum.de (visited on May 30, 2010)

15. Koch, N., Kraus, A.: The Expressive Power of UML-based Web Engineering. In: Second International Workshop on Web-oriented Software Technology (IWWOST 2002), vol. 16. Citeseer (2002)
16. Dukaczewski, B.R.M., Reiss, D., Stein, M.: MontiWeb - Modular Development of Web Information Systems. In: Proceedings of the 9th OOPSLA Workshop on Domain-Specific Modeling (DSM 2009), Orlando, Florida, USA (2009)
17. Mod4j. Website. http://www.mod4j.org/ (visited on May 30, 2010)
18. Pohl, K., Böckle, G., van der Linden, F.J.: Software Product Line Engineering: Foundations, Principles, and Techniques. Springer, Berlin (2005)
19. Visser, E.: WebDSL: A case study in domain-specific language engineering. In: Lämmel, R., Visser, J., Saraiva, J. (eds.) Generative and Transformational Techniques in Software Engineering II. LNCS, vol. 5235, pp. 291–373. Springer, Heidelberg (2008)

iBLOB: Complex Object Management in Databases through Intelligent Binary Large Objects

Tao Chen, Arif Khan, Markus Schneider*, and Ganesh Viswanathan

Department of Computer & Information Science & Engineering
University of Florida
Gainesville, FL 32611, USA
{tachen,ahkhan,mschneid,gv1}@cise.ufl.edu

Abstract. New emerging applications including genomic, multimedia, and geo-spatial technologies have necessitated the handling of complex *application objects* that are highly structured, large, and of variable length. Currently, such objects are handled using filesystem formats like HDF and NetCDF as well as the XML and BLOB data types in databases. However, some of these approaches are very application specific and do not provide proper levels of data abstraction for the users. Others do not support random updates or cannot manage large volumes of structured data and provide their associated operations. In this paper, we propose a novel two-step solution to manage and query application objects within databases. First, we present a generalized conceptual framework to capture and validate the structure of application objects by means of a *type structure specification*. Second, we introduce a novel data type called *Intelligent Binary Large Object (iBLOB)* that leverages the traditional BLOB type in databases, preserves the structure of application objects, and provides smart query and update capabilities. The iBLOB framework generates a type structure specific application programming interface (API) that allows applications to easily access the components of complex application objects. This greatly simplifies the ease with which new type systems can be implemented inside traditional DBMS.

1 Introduction

Many fields in computer science are increasingly confronted with the problem of handling large, variable-length, highly structured, complex *application objects* and enabling their storage, retrieval, and update by application programs in a user-friendly, efficient, and high-level manner. Examples of such objects include biological sequence data, spatial data, spatiotemporal data, multimedia data, and image data, just to name a few. Traditional database management systems (DBMS) are well suited to store and manage large, unstructured alphanumeric data. However, storing and manipulating large, structured application objects at the low byte level as well as providing operations on them are hardly supported. *Binary large objects (BLOBs)* are the only means to store such objects. However, BLOBs represent them as low-level, binary strings and do not preserve their structure. As a result, this database solution turns out to be unsatisfactory.

* This work was partially supported by the National Aeronautics and Space Administration (NASA) under the grant number NASA-AIST-08-0081.

A. Dearle and R.V. Zicari (Eds.): ICOODB 2010, LNCS 6348, pp. 85–99, 2010.

Hence, scientists have designed special file formats like *NetCDF* (*network Common Data Form*) and *HDF5* (*Hierarchical Data Format*) to store such objects in files. Unfortunately, without the support of a DBMS, standard features like a query language, concurrency control, transaction management, security, and recovery are unavailable (*data management problem*). A widely accepted approach to handling complex data in databases is to model and implement them as values of *abstract data types* (*ADT*) in a type system, or *algebra*, which is then embedded into an extensible DBMS and its query language. This enables their use as attribute data types in a database schema without disclosing the implementation details of their complex internal structure to the user and DBMS. At the type system level, extensible DBMS enable the specification of new ADTs like spatial, image, and XML data types. However, these ADTs have DBMS specific implementations and are not universally deployable (*generality problem*). On the other hand, BLOBs are not well suited for structured object management. They have originally been designed for storing unstructured data as byte sequences and offer a low-level interface for simple read/write access to byte ranges. Thus BLOBs do not understand the semantics of the internal structure of the application objects stored in them and therefore do not include methods to access internal components of them (*abstraction problem*). This makes the access to a component of an application object rather expensive since the entire object needs to be loaded into main memory to understand its structural semantics and get access to the component of interest. Further, BLOBs typically allow data to be appended, truncated, and modified through the overwriting of bytes. However, general data insertions and deletions are not supported unless the user explicitly shifts data (*update problem*).

In this paper, we present a novel, generic model for complex object management that focuses on providing the required functionality to address the data management, generality, abstraction, and update problems. We first propose a generalized method, named *type structure specification*, for representing and interpreting the structure of application objects. This specification provides an interface for the ADT implementer to describe the structure of complex objects at the conceptual level. Based on this specification, we employ a generalized framework, called *intelligent binary large objects* (*iBLOBs*), for the efficient and high-level storage, retrieval, and update of hierarchically structured complex objects in databases. iBLOBs store complex objects by utilizing the unstructured storage capabilities of DBMS and provide component-wise access to them. In this sense, they serve as a communication bridge between the high-level abstract type system and the low-level binary storage. This framework is based on two orthogonal concepts called *structured index* and *sequence index*. A *structured index* facilitates the preservation of the structural composition of application objects in unstructured BLOB storage. A *sequence index* is a mechanism that permits full support of *random updates* in a BLOB environment.

Section 2 describes relevant research related to the iBLOB concept. In Section 3, we describe the applications that involve large structured application objects, the existing approaches to handling them, and our approach to dealing with structured objects in a database context. We introduce the concept of type structure specification and the iBLOB framework in Sections 4 and 5. Finally, in Section 6, we draw some conclusions and discuss future work.

2   Related Work

The need for extensibility in databases, in general, and for new data types in databases [11], in particular, has been the topic of extensive research from the late eighties. In this section, we review work related to the storage and management of structured large application objects. The four main approaches can be subdivided into *specialized file formats*, new *DBMS prototypes*, *traditional relational DBMS*, and *object-oriented extensibility mechanisms in DBMS*.

The *specialized file formats* can be further categorized into text formats and binary formats [8]. Text formats organize data as a stream of Unicode characters whereas binary formats store numbers in "native" formats. XML [4] is a universal standard text data format primarily meant for data exchange. A critical issue with all text data formats is that they make the data structure visible and that one cannot randomly access specific subcomponent data in the middle of the file. The whole XML file has to be loaded into the main memory to extract the data portion of interest. Moreover, the methods used to define the legal structure for a XML document such as Document Type Definition (DTD) and XML Schema Definition (XSD) have several shortcomings. DTD lacks support for datatypes and inheritance, while XSD is really over-verbose and unintuitive when defining complex hierarchical objects. On the other hand, binary data formats like NetCDF [8,9] and HDF [1,8] support random access of subcomponent data. But updating an existing structure is not explicitly supported in both formats. Further, since HDF stores a large amount of internal structural specifications, the size of a HDF file is considerably larger than a flat storage format. Further, these file formats do not benefit from DBMS properties such as transactions, concurrency control, and recovery.

The second approach to storing large objects is the development of new *DBMS prototypes* as standalone data management solutions. These include systems such as *BSSS* [7], *DASDBS*, [10], *EOS* [3], *Exodus* [5], *Genesis* [2], and *Starburst* [6]. These systems operate on variable-length, uninterpreted byte sequences and offer low-level byte range operations for insertion, deletion, and modification. However, these systems do not manage structural information of large application objects and are hence unable to provide random access to object components.

The third approach taken to store large objects is the *use of tables and BLOBs* in traditional object-relational database management systems. Any hierarchical structure within an object can be incorporated in tables using a separate attribute column that cross-references tuples with their primary keys. Some database such as Oracle even support hierarchical SQL queries on such tables. However, the drawback of this method is that the querying becomes unintuitive and has to be supported by complex procedural language functions inside the database. Further, these queries are slow because of the need of multiple joins between tables. Binary Large OBjects (BLOBs) provide another means to store large objects in databases. However, this is a mechanism for storing unstructured, binary data. Hence, the entire BLOB has to be loaded into main memory each time for processing purposes.

The fourth approach to storing large objects is the use of *object-oriented extension mechanisms* in databases. Most popular DBMS support the CREATE TYPE construct to create user-defined data types. However, the type constructors provided (like array constructors) do not allow to create large *and* variable-length application objects.

3   Problems with Handling Structured Application Objects in Database Systems and Our Solution

Application objects like DNA structures, 3D buildings, and spatial regions are complex, highly structured, and of variable representation length. The desired operations on the application objects usually involve high complexity, long execution time and large memory. For example, *region* objects are complex application objects that are frequently used in GIS applications. As shown in Figure 1, a region object consists of components called *faces*, and *faces* are enclosed by *cycles*. Each *cycle* is a closed sequence of connected *segments*. Applications that deal with regions might be interested in numeric operations that compute the *area*, the *perimeter* and the *number of faces* of a region. They might also be interested in geometric operations that compute the *intersection*, *union*, and *difference* of two regions. Many more operations on regions are relevant to applications that work with maps and images. In any case, the implementation of an operation requires easy access to components of structured objects (e.g., segments, cycles, and faces of a region) that uses less memory and runs in less time.

Since database systems provide built-in advanced features like the SQL query language, transaction control, and security, handling complex objects in a database context is an expedient strategy. Most approaches are built upon two important architectures that enable database support for applications involving complex application objects.

Early approaches apply a *layered architecture* as shown in Figure 2a, in which a *middleware* that handles complex application objects is clearly separated from the *application front-end* that provides services and analysis methods to its users. In this architecture, only the underlying primitive data are physically stored in traditional RDBMS tables. The knowledge about the structure of complex objects is maintained in the middleware. It is the responsibility of the middleware to load the primitive data from the underlying database tables, to reconstruct complex objects from the primitive data, and to provide operations on complex objects. The underlying DBMS in the layered architecture does not understand the semantics of the complex data stored. In this sense, the database is of limited value, and the burden is on the application developer to implement a middleware for handling complex objects. This complicates and slows down the application development process.

A largely accepted approach is to model and implement complex data as *abstract data types* (*ADTs*) in a type system, or *algebra*, which is then embeded into an

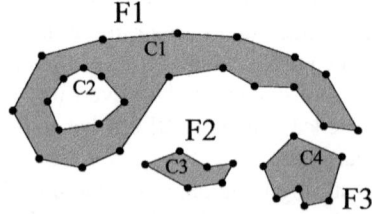

Fig. 1. A region object as an example of a complex, structured application object. It contains the faces F1, F2, and F3, which consist of the cycles C1 and C2 for F1, C3 for F2, and C4 for F3.

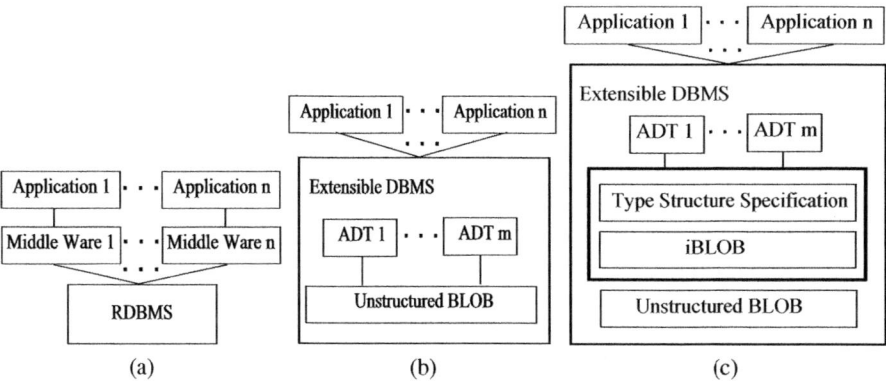

Fig. 2. The layered architecture (a) and the integrated architecture (b) and our solution (c)

extensible DBMS and its query language. This approach employs an *integrated architecture* (Figure 2(b)), where the applications directly interact with the extended database system, and use the ADTs as attribute data types in a database schema. Some commercial database vendors like Oracle and Postgres have included some ADTs like spatial data types as built-in data types in their database products. Extensible DBMS provides users the interfaces for implementing their own ADT so that all types of applications can be supported. Since the only available data structure for storing complex objects with variable length is BLOB, the implementations of ADTs for complex objects are generally based on BLOBs. The implementation of an abstract data type involves three tasks, the design of binary representation, the implementation of component retrieval and update, and the implementation of high level operations and predicates.

The integrated architecture has obvious advantages. It transfers the burden of handling complex objects from the application developer to databases. Once abstract data types are designed and integrated into a database context, applications that deal with complex objects become standard database applications, which require no special treatment. This simplifies and speeds up the development process for complex applications. However, the drawback of this approach is that ADTs for structured application objects rely on the unstructured BLOB type, which provides only byte level operations that complicate, or even foil, the implementation of component retrieval and update. Byte manipulation is a redundant and tedious task for *type system implementers* who want to implement a high-level type system because they want to focus on the design of the data types and the algorithms for the high-level operations and predicates.

In this paper, we propose a new concept that extends the integrated architecture approach, provides the type system implementers with a high level access to complex objects, and is capable of handling any structured application objects. In our concept, we apply the integrated architecture approach and extend it with a generalized framework (Figure 2c) that consists of two components, the *type structure specification* (Section 4) and the *intelligent BLOB* concept (Section 5). The type structure specification consists of algebraic expressions that are used by type system implementers to specify the internal hierarchy of the abstract data type. It is later used as the meta data for the intelligent

BLOB to identify the semantic meaning of each structure component. Further, as part of the type structure specification we provide a set of high-level functions as interfaces for type system implementers to create, access, or manipulate data at the component level. To support the corresponding interfaces, we propose a generic storage method called *intelligent BLOB (iBLOB)*, which is a binary array whose implementation is based on the BLOB type and which maintains hierarchical information. It is "intelligent" because, unlike BLOBs, it understands the structure of the object stored and supports fast access, insertion and update to components at any level in the object hierarchy.

The type structure specification in the framework provides an abstract view of the application object which hides the implementation details of the underlying data structure. The underlying intelligent BLOBs ensure a generic storage solution for any kinds of structured application objects, and enable the implementation of the high-level interfaces provided by the type structure specification. Therefore, the type structure specification and the concept of intelligent BLOBs together enable an easy implementation for abstract data types. type system implementers can be released from the task of interpreting the logical semantics of binary unstructured data, and the component level access is natively supported by the underlying iBLOB.

4 Representing and Interpreting Structured Application Objects with Type Structure Specifications

The structures of different application objects can vary. Examples are the structure of a region (Figure 1) and the structure of a book. We aim at developing a generic platform that accommodates all kinds of hierarchical structures. Thus, the first step is to explore and extract the common properties of all structured objects. Unsurprisingly, the hierarchy of a structured object can always be represented as a tree. Figure 3a shows the tree structure of a *region* object. In the figure, *face[]*, *holeCycle[]*, and *segment[]* represent a list of faces, a list of hole cycles and a list of segments respectively. In the tree representation, the root node represents the structured object itself, and each child node represents a component named *sub-object*. A sub-object can further have a structure, which is represented in a sub-tree rooted with that sub-object node. For example, a region object in Figure 3a consists of a label component and a list of face components. Each face in the face list is also a structured object that contains a face label, an outer cycle, and a list of hole cycles, where both the outer cycle and the hole cycles are formed by segments lists. Similarly, the structure of a book can also be represented as a tree (Figure 3b).

Further, we observe that two types of sub-objects can be distinguished called *structured objects* and *base objects*. Structured objects consist of sub-objects, and base objects are the smallest units that have no further inner structure. In a tree representation, each leaf node is a base object while internal nodes represent structured objects.

A tree representation is a useful tool to describe hierarchical information at a conceptual level. However, to give a more precise description and to make it understandable to computers, a formal specification would be more appropriate. Therefore, we propose a generic *type structure specification* as an alternative of the tree representation for describing the hierarchical structure of application objects.

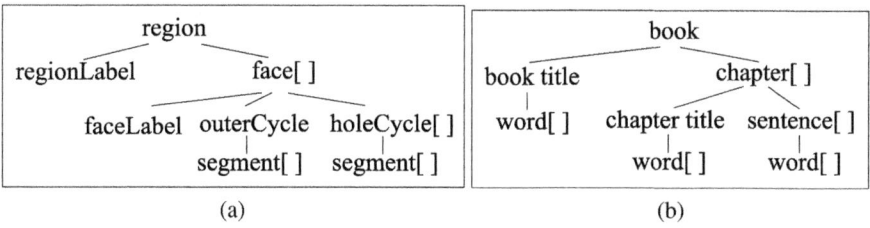

(a)   (b)

Fig. 3. The hierarchical structure of a *region* object and the hierarchical structure of a *book* object

We first introduce the concept of *structure expressions*. Structure expressions define the hierarchy of a structured object. A structure expression is composed of *structure tags (TAGs)* and *structure tag lists (TAGLISTs)*. A structure tag (TAG) provides the *declaration* for a single component of a structured object, whereas a structure tag list (TAGLIST) provides the declaration for a list of components that have the same structure. The declaration of a TAG, named *tag declaration*, is $\langle NAME : TYPE \rangle$, where *NAME* is the identifier of the tag and the value of *TYPE* is either *SO*, which is a flag that indicates a structured object, or *BO*, which is a flag that indicates a base object. An example of a *structured object tag* is $\langle region : SO \rangle$, and $\langle segment : BO \rangle$ is an example of a *base object tag*. We first define a set of terminals that will be used in structure expressions as constants. Then, we show the syntax of structure expressions.

Terminal Set $S = \{:=, \langle, \rangle, |, [,], SO, BO, :\}$

Expression ::= TAG := $\langle TAG \mid TAGLIST \rangle^{+}$;
TAGLIST ::= TAG[ ]
TAG   ::= $\langle NAME : TYPE \rangle$
TYPE   ::= $\langle SO \mid BO \rangle$
NAME   ::= IDENTIFIER

In the region example, we can define the structure of a region object with the following expression: $\langle region : SO \rangle := \langle regionLabel : BO \rangle \langle face : SO \rangle [ ]$. In the expression, the left side of := gives the tag declaration of a region object and the right side of := gives the tag declarations of its components, in this case, the region label and the face list. Thus, we say the region object is *defined* by this structure expression.

With structure expressions, the type system implementer can recursively define the structure of structured sub-objects until no structured sub-objects are left undefined. A list of structure expressions then forms a specification. We call a specification that consists of structure expressions and is organized following some rules a *type structure specification (TSS)* for an abstract data type. Three rules are designed to ensure the *correctness* and *completeness* of a type structure specification when writing structure expressions: (1) the first structure expression in a TSS must be the expression that defines the abstract data type itself (*correctness*); (2) every structured object in a TSS has to be defined with one and only one structure expression (*completeness* and *uniqueness*); (3) none of the base objects in a TSS is defined (*correctness*). By following these rules, the type system implementer can write one type structure specification for each abstract data type. Further, it is not difficult to observe that the conversion between a tree representation and a type structure specification is simple. The root node in a tree maps to the first structure expression in the TSS. Since all internal nodes are structured sub-objects and leaf nodes are base sub-objects, each internal node has exactly one corresponding structure expression in the TSS, and leaf nodes require no structure expressions. The type structure specification of the abstract data type *region* corresponding to the tree structure in Figure 3a is as follows:

$$
\begin{aligned}
\langle region : SO\rangle &:= \langle regionLabel : BO\rangle\langle face : SO\rangle[\,]; \\
\langle face : SO\rangle &:= \langle faceLabel : BO\rangle\langle outerCycle : SO\rangle\langle holeCycle : SO\rangle[\,]; \\
\langle outerCycle : SO\rangle &:= \langle segment : BO\rangle[\,]; \\
\langle holeCycle : SO\rangle &:= \langle segment : BO\rangle[\,];
\end{aligned}
$$

The next step after specifying the structure is to create and store the application object into the database. The TSS provides a workable interface for the type system implementer to create, access and navigate through the object. This higher-level interface is the abstraction of the iBLOB interface. This abstraction along with the specification, frees the type system implementer from understanding the underlying data type iBLOB that is used for finally representing the application object in the database. Navigating through the structure of the object is done by specifying a path from the root to the node by a string using the *dot-notation*. For example, to point to the first segment of the outer cycle of the third face of a region object can be specified by the string *region.face*[3].*outerCycle.segment*[1]. A component number (e.g., *first* segment, *third* face) is determined by the temporal order when a component was inserted. An important point to mention is that the structural validity of a path (e.g., whether an outer cycle is a subcomponent of a face) can be verified by parsing the TSS. However, the existence of a third face can only be detected during runtime. The set of operators which are defined by the interface are given below:

$$
\begin{aligned}
create &: \to SO \\
get &: path \to BO[\,] \\
set &: path \to bool \\
set &: path \times char* \to bool \\
baseObjectCount &: path \to int \\
subObjectCount &: path \to int
\end{aligned}
$$

An application object can be created by the operator *create*() which generates an empty *application object*. The operator *get*(*p*) returns all base objects at leaf nodes under the node specified by any valid path *p*. Since no data types are defined for the structured objects in intermediate nodes, these objects are not accessible, and paths to them are undefined. Hence, paths to intermediate nodes are interpreted differently in the sense that the operator *get*(*p*) recursively identifies and returns all base objects under *p*. The operator *set*(*p*) creates an intermediate component. The operator *set*(*p*, *s*) inserts a base object given as a character string *s* at the location specified by the path *p*. The last two operators *baseObjectCount*(*p*) and *subObjectCount*(*p*) return the number of base objects and the number of sub-objects under a node specified by the path *p*. As an example, for a region object with one face that contains an outer cycle with three segments, the corresponding code for creating the region object is given below:

region r = create(); *r.set*(*region.regionLabel,"MyRegion"*);
r.set(*region.face*[1]); *r.set*(*region.face*[1].*faceLabel,"Face1"*);
r.set(*region.face*[1].*outerCycle*);
r.set(*region.face*[1].*outerCycle.segment*[1],*seg1*);
r.set(*region.face*[1].*outerCycle.segment*[2],*seg2*);
r.set(*region.face*[1].*outerCycle.segment*[2],*seg3*);

The first line of the code shows how the type system implementer can create a region object based on the specified type structure specification. The second line creates the first face and the third line its outer cycle as intermediate components. The following three lines store the three segments *seg1*, *seg2*, *seg3* as components of the outer cycle.

5   Intelligent Binary Large Objects (iBLOBs)

In this section, we present the conceptual framework for a new database data type called *iBLOB* for *Intelligent Binary Large Objects*. This type enhances the functionality of traditional binary large objects (BLOBs) in database systems. Our concept also helps to solve the generality, abstraction and update problems (described in Section 1) that are exhibited by current approaches (see Section 2) to manage large application objects. BLOBs serve currently as the only means to store large objects in DBMS. However, they do not preserve the structure of application objects and do not provide access, update and query functionality for the sub-components of large objects. *iBLOBs* help to smartly extend traditional BLOBs by preserving the object structure internally and providing application-friendly access interfaces to the object components. All this is achieved while maintaining low level access to data and extending existing database systems using object-oriented constructs and *abstract data types* (*ADTs*).

The iBLOB framework consists of two main sections called the *structure index* and the *sequence index* (Figure 4). The first section contains the *structure index* which helps us represent the object structure as well as the base data. The second section contains the *sequence index* that dictates the sequential organization of object fragments and preserves it under updates. Since the underlying storage structure of an iBLOB is provided through a BLOB, which is available in most DBMSs, the iBLOB data type can be registered as a user-defined data type and be used in SQL.

5.1   iBLOB Structure Index: Preserving Structure in Unstructured Storage

A structure index is a mechanism that allows an arbitrary hierarchical structure to be represented and stored in an unstructured storage medium. It consists of two components for, first, the representation of the structure of the data and, second, the actual data

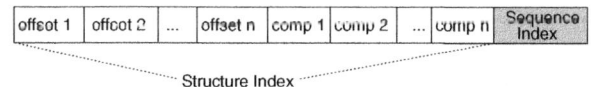

Fig. 4. Illustration of an iBLOB object consisting of a structure index and a sequence index

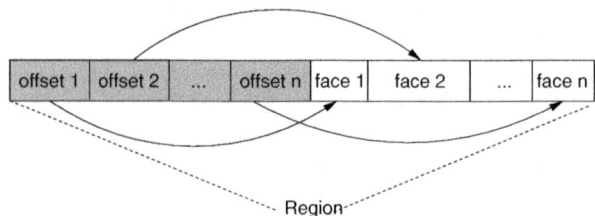

Fig. 5. A structured object consisting of n sub-objects and n internal offsets

themselves. The structural component is used as a reference to access the data's structural hierarchy. The mechanism is not intended to enforce constraints on the data within it; thus, it has no knowledge of the semantics of the data upon which it is imposed. This concept considers hierarchically structured objects as consisting of a number of variable-length sub-objects where each sub-object can either be a *structured object* or a *base object*. Within each structured object, its sub-objects reside in sequentially numbered slots. The leaves of the structure hierarchy contain base objects.

To illustrate the concept of a structure index, we show an example how to store a spatial region object with a specific structure in a database. A region data type may be described by a hierarchical structure as shown in Figure 3a. Consider a region made up of several faces. If we needed to access the 50th face of a region object using a traditional BLOB storage mechanism, one would have to load and sequentially traverse the entire BLOB until the desired face would be found. Further, since the face objects can be of variable length, the location of the 50th face cannot be easily computed without extra support built in to the BLOB. In order to avoid an undesirable sequential traversal of the BLOB, we introduce the notion of *offsets* to describe structure. Each hierarchical level of a structure in a structure index stored in a BLOB is made up of two components (corresponding to the two components of the general structure index described above). The first component contains offsets that represent the location of specific sub-objects. The second component represents the sub-objects themselves. We define offsets to have a fixed size; thus, the location of the ith face can be directly determined by first calculating the location of the ith offset and then reading the offset to find the location of the face. Figure 5 shows a structured object with internal offsets.

The recursive nature of hierarchical structures allows us to generalize the above description. Each sub-object can itself have a structure like the region described above. Objects at the same level are not required to have the same structure; thus, at any given level it is possible to find both structured sub-objects and base objects (raw data). For example, we can extend the structure of a region object so that it is made up of a collection of faces each of which contains an outer cycle and zero or more hole cycles, which in turn are made up of a collection of segments. Segments can be implemented as a pair of (x, y)-coordinate values. This example is illustrated in terms of structured and base objects in Figure 6 where the top level object represents a region with an information part, a label, and one of its face sub-objects.

In general, a specific structure index implementation must be defined with respect to the underlying unstructured storage medium. Because we have to use BLOBs as

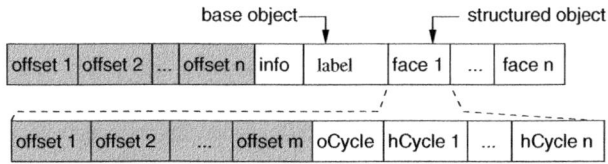

Fig. 6. A structured object consisting of a base object and structured sub-objects

the only alternative in a database context, we are forced to embed both the structure index and the data into a BLOB. Thus, using an offset structure embedded within the data itself is an appropriate solution. However, this may not be ideal in all cases. For instance, one could implement a structure index for data stored in flat files. In this case, the structure index and the data could be represented in seperate files. In general, the structure index concept must be adapted to the capabilities of the user's desired storage medium for implementation.

5.2   iBLOB Sequence Index: Tracking Data Order for Updates

Different DBMSs provide different implementations of the BLOB type with varied functionalities. However, most advanced BLOB implementations support three operations at the byte level, namely, random read and append (write bytes at end of BLOB), truncate (delete bytes at end) and overwrite (replace bytes with another block of bytes of the same or smaller length).

Structured large objects require the ability to update sub-objects within a structure. Specifically, they require *random updates* which include insertion, deletion and the ability to replace data with new data of arbitrary size. Examples are the replacement of a segment by several segments in a cycle of a region object, or adding a new face. Given a large region object, updating it entirely for each change in a face, cycle or segment becomes very costly when stored in BLOBs (update problem). Thus, it is desirable to update only the part of the structure that needs updating. For this purpose, we present a novel *sequence index* concept that is based on the random read and data append operations supported by BLOBs Extra capabilities provided by higher level BLOBs are a further improvement and serve for optimization purposes. The sequence index concept is based on the idea of physically storing new data at the end of a BLOB and providing an index that preserves the logically correct order of data.

Consequently, data will have internal fragmentation and will be physically stored out-of-order, as illustrated in Figure 7. In this figure, the data blocks (with start and end

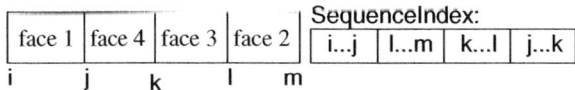

Fig. 7. An out-of-order set of data blocks and their corresponding sequence index

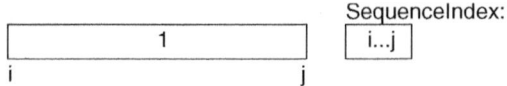

Fig. 8. The initial in-order and defragmented data and sequence index

byte addresses represented by letters under each boundary) representing faces should be read in the order $1, 2, 3, 4$, even though physically they are stored out-of-order in the BLOB (we will study the possible reasons shortly). By using an ordered list of physical byte address ranges, the sequence index specifies the order in which the data should be read for sequential access. The sequence index from Figure 7 indicates that the block $[i \ldots j]$ must be read first, followed by the block $[l \ldots m]$, etc.

Based on the general description of the sequence index given above, we now show how to apply it as a solution to the update problem. Assume that the data for a given structured object is initially stored sequentially in a BLOB, as shown in Figure 8. Suppose further that the user then makes an insertion at position k in the middle of the object. Instead of shifting data after position k within the BLOB to make room for the new data, we append it to the BLOB as block $[j \ldots l]$, as shown in Figure 9. By modifying the sequence index to reflect the insertion, we are able to locate the new data at its logical position in the object.

Figure 10 illustrates the behavior of the sequence index when a block is intended to be deleted from the structured object. Even though there is no new data to append to the BLOB, the sequence index must be updated to reflect the new logical sequence. Because the BLOB does not actually allow for the deletion of data, the sequence index is modified in order to prevent access to the deleted block $[m \ldots n]$ of data. This can result in internal fragmentation of data in the iBLOB which can be managed using a special *resequence* operation shown later in the iBLOB interface.

Finally, Figure 11 illustrates the case of an update where the values of a block of data $[o \ldots p]$ as a portion of block $[j \ldots l]$ are replaced with values from a new block $[l \ldots q]$. For this kind of update, it is possible for the new set of values to generate a block size different from that of the original block being replaced.

iBLOBs enhance BLOBs by providing support for truncate and overwrite operations at the higher *component level* of an application object's structure. The *truncate* operation in BLOB (delete bytes at end) is enhanced in iBLOB with a *remove* function which can perform deletion of components at any location (beginning, middle or at the end of structure) as shown in Figure 10. The *overwrite* operation in BLOB (replace byte array with another of same length) is enhanced in iBLOB with a combination of

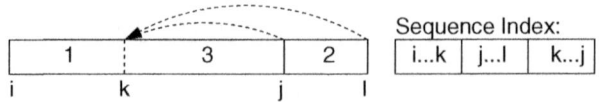

Fig. 9. A sequence index after inserting block $[j \ldots l]$ at position k

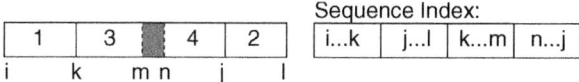

Fig. 10. A sequence index after deleting block $[m\ldots n]$

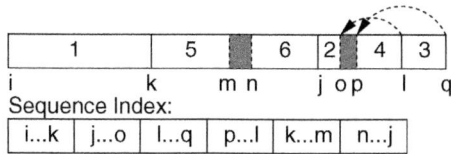

Fig. 11. A sequence index after replacing block $[o\ldots p]$ by block $[l\ldots q]$

remove and *insert* functions and sequence index adjustments, to perform the overwrite of components with other components of different sizes as shown in Figure 10.

5.3 The iBLOB Interface

In this section, we present a generic interface for constructing, retrieving and manipulating iBLOBs. Within this *iBLOB interface*, we assume the existence of the following data types: the primitive type *Int* for representing integers, *Storage* as a storage structure handle type (i.e., blob handle, file descriptor, etc.), *Locator* as a reference type for an iBLOB or any of its sub-objects, *Stream* as an output channel for reading byte blocks of arbitrary size from an iBLOB object or any of its sub-objects, and *data* as a representation of a base object. Figure 12 lists the operations offered by the interface. We use the term *l-referenced object* to indicate the object that is referred to by a given locator *l*. The following descriptions for these operations are organized by their functionality:

- **Construction and Duplication:** An iBLOB object can be constructed in three different ways. The first constructor *create*() (1) creates an empty iBLOB object. The second constructor *create*(*sh*) (2) constructs an iBLOB object from a specific storage structure handle *sh* such as a BLOB object handle or a file descriptor. The third constructor *create*(*s*) (3) is a copy constructor and builds a new iBLOB object from an existing iBLOB object *s*. Similarly, an iBLOB object s_2 can also be copied into another iBLOB object s_1 by using the *copy*(s_1, s_2) operator (4).
- **Internal Reference:** In order to provide access to an internal sub-object of an iBLOB object, we need a way to obtain the reference of such a sub-object. The sub-object referencing process must start from the topmost hierarchical level of the iBLOB object *s* whose locator *l* is provided by the operator *locateiBLOB*(*s*) (5). From this locator *l*, a next level sub-object can be referenced by its slot *i* in the operator *locate*(*s, l, i*) (6).
- **Read and Write:** Since iBLOBs support large objects which may not fit into main memory, we provide a stream based mechanism through the operator *getStream*(*s, l*) (7) to consecutively read arbitrary size data from any l-referenced object. The

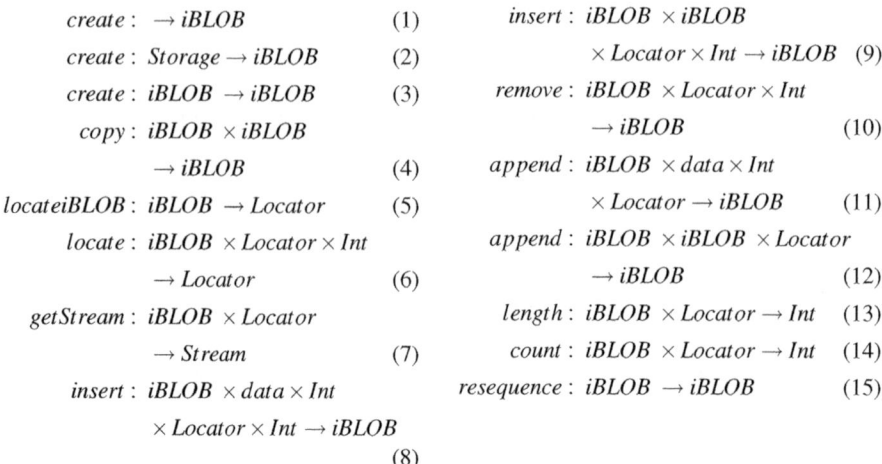

$$create: \ \rightarrow iBLOB \qquad (1)$$
$$create: \ Storage \rightarrow iBLOB \qquad (2)$$
$$create: \ iBLOB \ \rightarrow iBLOB \qquad (3)$$
$$copy: \ iBLOB \ \times iBLOB$$
$$\rightarrow iBLOB \qquad (4)$$
$$locateiBLOB: \ iBLOB \ \rightarrow Locator \qquad (5)$$
$$locate: \ iBLOB \ \times Locator \times Int$$
$$\rightarrow Locator \qquad (6)$$
$$getStream: \ iBLOB \ \times Locator$$
$$\rightarrow Stream \qquad (7)$$
$$insert: \ iBLOB \ \times data \times Int$$
$$\times Locator \times Int \rightarrow iBLOB \qquad (8)$$

$$insert: \ iBLOB \ \times iBLOB$$
$$\times Locator \times Int \rightarrow iBLOB \qquad (9)$$
$$remove: \ iBLOB \ \times Locator \times Int$$
$$\rightarrow iBLOB \qquad (10)$$
$$append: \ iBLOB \ \times data \times Int$$
$$\times Locator \rightarrow iBLOB \qquad (11)$$
$$append: \ iBLOB \ \times iBLOB \ \times Locator$$
$$\rightarrow iBLOB \qquad (12)$$
$$length: \ iBLOB \ \times Locator \rightarrow Int \qquad (13)$$
$$count: \ iBLOB \ \times Locator \rightarrow Int \qquad (14)$$
$$resequence: \ iBLOB \ \rightarrow iBLOB \qquad (15)$$

Fig. 12. The standardized iBLOB interface

stream obtained from this operator behaves similarly to a common file output stream. Other than reading data, the interface allows insertion of either a base object d of specified size z through the operator $insert(s,d,z,l,i)$ (8) or an entire iBLOB object s_1 through the operator $insert(s,s_1,l,i)$ (9) into any l-referenced object at a specified slot i. A base object d such as in the operator $append(s,d,z,l)$ (11) or a iBLOB object s_1 such as in operator $append(s,s_1,l)$ (12) can be appended to an l-referenced object. This is effectively the same as inserting the input as the last sub-object of the referenced object. The operator $remove(s,l,i)$ (10) removes the sub-object at slot i from the parent component with Locator l.

– **Properties and Maintenance:** The actual size of an l-referenced object is obtained by using the operator $length(s,l)$ (13) while the number of sub-objects of the object is provided by the operator $count(s,l)$ (14). Finally, the operator $resequence(s)$ (15) reorganizes and defragments the iBLOB object s collapsing its sequence index such that it contains a single range. This operation effectively synchronizes the physical and logical representations of the iBLOB object and minimizes the storage space.

To test the functionality we have implemented the iBLOB data type in Oracle, Informix and PostgreSQL using object oriented extensions and programming API of the DBMS. Due to space constraints, we have omitted the iBLOB implementation details in this paper. Each operator in the TSS interface can be implemented using the corresponding iBLOB interface operator. For e.g., to implement $get(region.face[1].outerCycle.segment[1])$, we first use $locateiBLOB$ (5) to get a Locator to the iBLOB, then use $locate()$ (5) repeatedly to move across levels and navigate to the required component (i.e., first segment), and finally, $getStream()$ (7) to retrieve the first segment of the outerCycle in fifth face. Other TSS interface functions like set, $baseObjectCount$ and $subObjectCount$ can also be implemented in a similar manner.

6   Conclusions

In this paper, we provide a novel solution to store and manage complex application objects (i.e., variable length, structured, hierarchical data) by introducing a new mechanism for handling structured objects inside DBMSs. This includes two major concepts. First, we present a *type structure specification* (*TSS*) that helps to describe the structure of complex application objects. Then we introduce a special SQL data type called *Intelligent Binary Large Object* or *iBLOB* that enables the database to handle structured objects. The combination of type structure specification and iBLOBs provides the necessary tools to easily implement type systems in databases. However, the focus of this paper is to extend database functionality to natively support complex objects. As future work, we plan to optimize iBLOBs for performance.

References

1. HDF-Hierarchical Data Format, http://www.hdfgroup.org/
2. Batory, D.S., Barnett, J.R., Garza, J.F., Smith, K.P., Tsukuda, K., Twichell, B.C., Wise, T.E.: Genesis: an Extensible Database Management System. IEEE Trans. on Software Engineering 14, 1711–1730 (1988)
3. Biliris, A.: The Performance of Three Database Storage Structures for Managing Large Objects. In: ACM SIGMOD Int. Conf. on Management of Data, pp. 276–285 (1992)
4. Bray, T., Paoli, J., Sperberg-McQueen, C.M., Maler, E., Yergeau, F.: Extensible markup language (XML) 1.0. W3C recommendation, 6 (2000)
5. Carey, M.J., DeWitt, D.J., Vandenberg, S.L.: A Data Model and Query Language for Exodus. ACM SIGMOD Record 17, 413–423 (1988)
6. Haas, L.M., Chang, W., Lohman, G.M., McPherson, J., Wilms, P.F., Lapis, G., Lindsay, B.G., Pirahesh, H., Carey, M.J., Shekita, E.J.: Starburst Mid-flight: As the Dust Clears. IEEE Trans. on Knowledge and Data Engineering (TKDE) 2, 143–160 (1990)
7. Hwang, B., Jung, I., Moon, S.: Efficient storage management for large dynamic objects. In: EUROMICRO 1994, System Architecture and Integration 20th EUROMICRO Conference, pp. 37–44 (September 1994)
8. McGrath, R.E.: XML and Scientific File Formats. The Geological Society of America (2003)
9. Rew, R.K., Ucar, B., Hartnett, E.J.: Merging netCDF and HDF5. In: 20th Int. Conf. on Interactive Information and Processing Systems (2004)
10. Schek, H.-J., Paul, H.-B., Scholl, M.H., Weikum, G.: The DASDBS Project: Objectives, Experiences, and Future Prospects. IEEE Trans. on Knowledge and Data Engineering (TKDE) 2(1), 25–43 (1990)
11. Stonebraker, M.: Inclusion of New Types in Relational Data Base Systems. Int. Conf. on Data Engineering Conference (ICDE), pp. 262–269 (1986)

Object-Oriented Constraints for XML Schema

Suad Alagić[1], Philip A. Bernstein[2] and Ruchi Jairath[1]

[1] Department of Computer Science,
University of Southern Maine
alagic@usm.maine.edu, ruchi.jairath@maine.edu
[2] Microsoft Research
philbe@microsoft.com

Abstract. This paper presents an object-oriented representation of the core structural and constraint-related features of XML Schema. The structural features are represented within the limitations of object-oriented type systems including particles (elements and groups) and type hierarchies (simple and complex types and type derivations). The applicability of the developed representation is demonstrated through a collection of complex object-oriented queries. The main novelty is that features of XML Schema that are not expressible in object-oriented type systems such as range constraints, keys and referential integrity, and type derivation by restriction are specified in an object-oriented assertion language Spec#. An assertion language overcomes major problems in the object-oriented/XML mismatch. It allows specification of schema integrity constraints and transactions that are required to preserve those constraints. Most importantly, Spec# technology comes with automatic static verification of code with respect to the specified constraints. This technology is applied in the paper to transaction verification.

1   Introduction

XML Schema Definition language (XSD for short) is a widely-used standard for specifying structural features of XML data [18]. In addition, XSD allows specification of constraints that XML data is required to satisfy. But producing an object-oriented schema that reflects correctly the source XSD schema and adheres to the type systems of mainstream object-oriented languages presents a major challenge [9]. Such an object-oriented interface is required by database designers, users writing queries and transactions, and application programmers in general when processing XML data that conforms to XSD.

There are two broad types of features in XSD: structural and constraint-based. The structural features are represented by the features of the type system. This includes elements, attributes, groups, and simple and complex types. Typical XSD constraints are range constraints that specify the minimum and maximum number of repeated occurrences, rules for type derivation by restriction that restrict the set of valid instances of a type, and identity constraints that define keys and referential integrity. Unfortunately, object-oriented type systems have severe limitations in representing these XSD constraints.

A. Dearle and R.V. Zicari (Eds.): ICOODB 2010, LNCS 6348, pp. 100–117, 2010.

Most of the existing object-oriented interfaces to XSD exhibit a number of problems due to the mismatch of XML and object-oriented type systems [7,8,11,12,19,20]. A more detailed account of these issues as they apply to specific approaches is given in [3]. Here we just mention some of these problems:

- Not distinguishing between elements and attributes in the object-oriented representation or not representing attributes at all.
- Not being able to represent repetition of elements and attributes with identical names (tags).
- Failing to represent correctly the particle structure of XSD (with elements and groups) and the range of occurrences constraints in particular.
- Confusing the particle hierarchy (with elements and groups) and the type hierarchy (with simple and complex types and type derivations) of XSD.
- Not distinguishing different types of XSD groups in the object-oriented representation (sequence versus choice) or not representing groups at all.
- In the object-oriented representation, not distinguishing the two type derivation techniques in XSD: by restriction and by extension.
- Failing to represent accurately XSD type derivation by restriction, facets and range constraints in particular.
- Having no representation of the XSD identity constraints (keys and referential integrity) and thus no way of enforcing them.

The key question is whether it is possible to develop an object-oriented interface that captures the core XSD structural features while avoiding at least some of the above problems. The main problem lies in the complexity of XSD, its semantics, and its mismatch with features of object-oriented type systems.

Our research contributions are as follows:

- We isolate the structural core of XSD which contains the essential structural features of XSD and abstracts away a variety of other XSD features [2,3].
- We demonstrate the utility of our interfaces by giving a variety of typical object-oriented queries specified in LINQ [10].
- We specify the object-oriented meta-level which consists of a full representation of features of the XSD core including particle structures (elements and groups), types and type derivations, content models and identity constraints).
- We specify XML Schema constraints in an object-oriented assertion language Spec#.
- We show how object-oriented schemas and transactions could be specified using general object-oriented constraints.
- We show how transactions are verified using static automatic verification in Spec#.

Due to the limitations of object-oriented type systems, we can represent only some of the constraints of the source XML Schema structurally. In the generated object-oriented schemas, range constraints are present as minOccurs and maxOccurs methods that return the bounds. The distinction in the semantics of different types of groups is represented by different interfaces whose default implementation is required to support different semantics. Type derivation by restriction is represented using not only inheritance, but also a hierarchy of interfaces representing different types of facets and overriding minOccurs and maxOccurs methods. Full details of the XSD schema are represented at the meta level. The XSD identity constraints are represented at the meta level by a hierarchy of interfaces representing different types of identity constraints. The same applies to the content models and the type derivation hierarchies.

While we can overcome some of the representation problems using a suitable structural representation, the key problem of the object-oriented/XML mismatch is that type systems cannot represent constraints expressible in XML Schema. This is why we use an object-oriented assertion language Spec# [13] which allows specification of range constraints, key and referential integrity constraints, and type derivation by restriction. In addition, more general application-oriented constraints not expressible in XML Schema or standard database technologies can now be specified and enforced using automatic verification that Spec# offers. Note that we always assume that the original XML Schema has been validated. For checking satisfaction of XML Schema constraints such as keys and referential integrity see [15]. In this research we consider object-oriented representation of those constraints. Automatic static verification of the object-oriented representation of XSD constraints is a major distinction with respect to our previous work [1,4,5] as well as with respect to other work [6,16,17].

The idea of static verification of transaction safety with respect to the database integrity constraints is not new [6,16,17] but it has never been implemented at a very practical level so that it can be used by typical database programmers. The first problem with object-oriented technology is that object-oriented schemas are not equipped with general integrity constraints, primarily because mainstream object-oriented languages do not have them. This problem gets resolved using an object-oriented assertion language such as JML, OCL or Spec#. Using an assertion language, schemas can now be specified with general database integrity constraints (invariants) and transactions can be specified in a declarative fashion with preconditions and postconditions.

However, the ability to verify statically that a transaction implemented in a mainstream object-oriented language satisfies those constraints has been out of reach. Our previous results [1,4,5] were based on a higher-order interactive verification system which is so sophisticated that it is unlikely to be used by database programmers. A pragmatic goal has been static automatic verification which hides completely the prover technology from users. With recent development in the Spec# technology this becomes possible. This is the main novelty of this paper.

Spec# has limitations in expressiveness dictated by the requirement for automatic static verification. We show that the range of Spec# features is surprisingly suitable for specification for XML Schema and other typical database integrity constraints. This specifically applies to existential and universal quantification required for key and referential constraints and Spec# comprehensions (sum, max, min, count etc.). The Spec# type system includes non-null object types and hence eliminates statically a very frequent error in application programs (and transactions) of trying to dereference a null pointer. Spec# allows explicit representation of aggregation of a complex object in terms of its components and constraints (invariants) that apply to a complex object and its components. Typically, a transaction does not maintain the required integrity constraints until its completion (commit). Spec# has a mechanism that allows specification of this situation so that it is properly handled by the verifier.

This paper is organized as follows. In sections 2, 3 and 5 we discuss the structural representation of the core of XML Schema within the limitations of object-oriented type systems. Sections 4, 6 and 7 deal with XML Schema constraints, their representation in Spec#, and their static automatic verification as it applies to transactions.

2   Object-Oriented Core of XML Schema

An XML document is a single element, which is the basic case of the XSD notion of a particle. The particle hierarchy contains a direct specification of the actual XML instances, which are documents. In general, a particle consists of a sequence of other particles, which may be elements or more general particles. The range of occurrences in a sequence is determined by invoking methods `minOccurs` and `maxOccurs`, but this range cannot be enforced by the type system. The default values of `minOccurs` and `maxOccurs` are both equal to 1.

```
interface XMLParticle
{ int minOccurs();
  int maxOccurs();
}
```

An element is a particular case of a particle. An element has a name (i.e., a tag) and a value. The value of an element may be simple or complex. The types of values of elements are structured into a separate hierarchy. If an element has a value of a complex type, that type contains the specification of the complex element structure.

```
interface XMLElement: XMLParticle
{ XMLName name();
  XMLanyType value();
}
```

Types of values of elements are structured into the type hierarchy specified below. The root of this type hierarchy is `XMLanyType`. An XML type may be simple or complex, hence the two immediate subtypes of `XMLanyType` are `XMLanySimpleType` and `XMLanyComplexType`.

```
    interface XMLanySimpleType: XMLanyType {...}
```

A value of an XML complex type in general consists of a set of attributes and a content model, where the latter is represented in this interface by its particle structure:

```
interface XMLanyComplexType:  XMLanyType {
   XMLSequence<XMLAttribute> attributes();
   XMLParticle particle();
}
```

XMLSequence is a parametric type whose implementation is C# IList. An attribute has a name (its tag) and a value. The value of an attribute is required to be simple, hence the following specification of an attribute type:

```
interface XMLAttribute  {
   XMLName name();
   XMLanySimpleType value();
}
```

A particle amounts to a sequence of terms. A term is either an element or a group. Since a range constraint may be associated with any type of a term, in a slightly simplified view, elements and groups are viewed as particles, which have range constraints. So we have:

```
interface XMLGroup: XMLParticle {
   XMLSequence<XMLParticle> particles();
}
```

There are three types of groups in XSD. Each of them is specified as a sequence of particles. For an all-group these particles must be elements. Hence the result of the method particles is covariantly overridden in the all-group. This is a situation that appears often and violates the typing rules for parametric types. This problem is circumvented somewhat below using the new feature of C# which amounts to hiding rather than overriding.

```
interface XMLSequenceGroup: XMLGroup {. . .}
interface XMLChoiceGroup: XMLGroup {. . .}
interface XMLAllGroup: XMLGroup {// . . .
   new XMLSequence<XMLElement> particles();
}
```

The semantics of sequence-group and choice-group are very different in XSD. An instance of a sequence-group is a sequence of particle instances. An instance of a choice-group contains just one of the particles specified in the choice-group. Specification of this semantic difference cannot be expressed in a satisfactory manner in an object-oriented type system alone [9]. It requires an assertion language. The underlying classes implementing the above interfaces have to correctly implement this semantics.

3   Object-Oriented Queries

In this section we illustrate the usage and suitability of the presented object-oriented interfaces to XSD by presenting a collection of object-oriented queries in the Language-Integrated Query (LINQ) feature of .NET [10]. The queries given below reflect complex group structure. AllJobOffers is an element whose type is a complex type JobOffers:

```
<xsd:element name = "AllJobOffers" type= "JobOffers" />
```

The particle structure of the type JobOffers is a sequence group. The first particle of this sequence-group is an element JobID. The second particle is a sequence-
group which consists of two elements: Name and SSN. This latter sequence-group
is repeated an unbounded but finite number of times, including zero times.

```
<xsd:complexType name = "JobOffers" >
  <xsd:sequence >
     <xsd:element name = "JobID" type = "xsd:string" />
      <xsd:sequence minOccurs="0" maxOccurs="unbounded"/>
              <xsd:element name = "Name" type = "xsd:string"/>
              <xsd:element name = "SSN" type = "xsd:int"/>
      </xsd:sequence>
  </xsd:sequence>
</xsd:complexType>
```

In the object-oriented representation AllJobOffers is an element whose value
has the type JobOffers.

```
interface AllJobOffers: XMLElement {
  new JobOffers value();
}
```

JobOffers is a complex type whose particle is of type JobSequence. JobSequence
is a sequence-group.

```
interface  JobOffers: XMLanyComplexType {
  new JobSequence particle();
}
interface JobSequence: XMLSequenceGroup {
   XMLString JobID();
   XMLSequence<JobGroup>  jobOffers();
   // set minOcurs and maxOccurs
}
```

XMLString and XMLInt are simple types derived from XMLanySimpleType and
represented by C# string and int types respectively. JobGroup is a sequence-
group whose particles are two elements: Name and SSN. The representation given
below is based on [2].

```
interface JobGroup: XMLSequenceGroup {
  new XMLsequence<XMLElement> particles();
  XMLString Name();
  XMLInt SSN(); }
```

An example of a LINQ query is given below:

```
static AllJobOffers J;
static JobSequence offers = J.value().particle();
IEnumerable<JobGroup> ProgrammingJobs =
               from x in offers.jobOffers()
               where offers.JobID() == "Programmer"
               select x;
```

To construct instances of a new type, the corresponding class must be defined first. Given a class

```
class AnOffer: XMLElement
{ AnOffer(XMLString name, XMLint salary){// . . .};
}
```

the query given below now makes use of the constructor in the above class for producing the output sequence of objects:

```
static AllJobOffers J;
static JobSequence G = J.value().particle();
  IEnumerable<AnOffer> ProgrammerOffer =
            from j in G.jobOffers()
            where G.JobID() == "Programmer"
            select (new AnOffer(G.JobID(), 100000));
```

4   Constraints in XML Schema

In this section we illustrate a variety of constraint related features of XML Schema that are not expressible in object-oriented type systems and hence require a strictly more powerful paradigm offered by object-oriented assertions languages. A range constraint is illustrated in the example below where the number of occurrences of a job offer is at least 1 and at most 100:

```
<xsd:complexType name ="JobOfferType" >
 <xsd:sequence>
     <xsd:element   name="JobID" type ="xsd:string" />
     <xsd:element   name="candidateName" type ="xsd:string" />
     <xsd:element   name ="SSN" type ="xsd:string"/>
 </xsd:sequence>
</xsd:complexType>
<xsd:complexType   name = "JobOffersType">
  <xsd:sequence>
    <xsd:element   name = "jobOffer"   type = "JobOfferType"
            minOccurs  = "1"   maxOccurs = "100" />
  </xsd:sequence>
</xsd:complexType>
```

Type derivation by restriction is illustrated below by a short list of job offers. The type ShortListedOffers is derived by restriction from the type JobOffersType by restricting the range of occurrences constraint in the type JobOffersType.

```
<xsd:complexType   name = "ShortListedOffers"
 <xsd:restriction base =  "JobOffersType"
   <xsd:sequence>
     <xsd:element name = "jobOffer"   type = "JobOfferType"
            minOccurs = "1"   maxOccurs = "10" />
   </xsd:sequence>
 </xsd:restriction>
</xsd:complexType>
```

The XML Schema style of specification of a key constraint is given below. This key is specified on the field JobID. The scope to which this key applies is specified by a selector which is a simplified XPath expression.

```
<xsd:complexType   name="JobType" >
 <xsd:sequence>
    <xsd:element   name ="JobID"   type="xsd:string"  />
    <xsd:element   name ="JobTitle" type ="xsd:string" />
    <xsd:element   name = "salary" type ="xsd:float" />
 </xsd:sequence>
</xsd:complexType>
<xsd:complexType name="JobsType" >
<xsd:sequence>
 <xsd:element name="job"   type = "JobType"
         minOccurs = "1"   maxOccurs ="1000" />
</xsd:sequence>
<xsd:key name = "JobIDkey" >
        <xsd:selector   xpath=" ./job" >
        <xsd:field xpath   ="JobID" >
</xsd:key>
</xsd:complexType>
```

The example below illustrates a key and referential integrity constraint. The key constraint specifies that JobID is a key in the sequence of job offers. The referential integrity constraint refers to the key constraint in the sequence of jobs. It requires that a job offer refers to an existing job in the sequence of jobs.

```
<xsd:complexType   name = "JobOffersType">
  <xsd:sequence>
    <xsd:element name = "jobOffer"   type = "JobOfferType"
                  minOccurs = "1"  maxOccurs = "100" />
 </xsd:sequence>
<xsd:key name="candidateKey" >
   <xsd:selector xpath = "./jobOffer"  />
   <xsd:field   xpath="JobID" />
</xsd:key>
<xsd:keyref   name = "JobRef" >
 <xsd:selector   xpath="JobsType/job" />
 <xsd:field xpath="JobID" />
</xsd:keyref>
</xsd:complexType>
```

5   Meta Level

The meta (schema) level contains as complete and accurate a representation of an XSD source schema as is possible within the framework of object-oriented type systems. Like in SOM [14] there exists an abstraction XMLSchemaObject so that all other schema object types are derived from it. A content model consists of a specification of a type and its type derivation:

```
interface XMLSchemaContentModel: XMLSchemaObject
{XMLSchemaType content();
 XMLSchemaTypeDerivation typeOfDerivation();
}
```

A content model may be simple or complex. If it is simple, the underlying type is simple and so is its type derivation:

```
interface XMLSchemaSimpleContent: XMLSchemaContentModel {
 new XMLSchemaSimpleType content();
 new XMLSchemaSimpleTypeDerivation  typeOfDerivation();
}
```

In the above interface the result types of both methods are overridden covariantly. Unlike Java, for some reason C# does not allow this type safe change of the signatures of inherited methods and hence requires hiding.

If a content model is complex, its underlying type may be either simple or complex. This is why the result type of the method content remains XMLSchemaType. If the underlying type is simple, the content model still may contain attributes. But if the content model is complex, the type derivation will be one of complex type derivations, as reflected in the result type of the method typeOfDerivation:

```
interface XMLSchemaComplexContent: XMLSchemaContentModel {
  XMLSchemaComplexTypeDerivation typeOfDerivation();
}
```

The interfaces that follow represent XSD type derivation rules. Every type derivation has a base type. If the type derivation is simple, the base type must be simple:

```
interface XMLSchemaTypeDerivation: XMLSchemaObject {
  XMLSchemaType base();
 }
interface XMLSchemaSimpleTypeDerivation: XMLSchemaTypeDerivation {
 new XMLSchemaSimpleType base();
}
```

There are two types of simple type derivation. Simple type derivation by restriction requires specification of a set of constraining facets. This structural representation is augmented in our approach using assertions as explained in section 6. Simple type extension allows only additional attributes:

```
interface XMLSimpleTypeRestriction: XMLSchema SimpleTypeDerivation {
  XMLSchemaSet<XMLFacet> facets();
}
interface XMLSchemaSimpleTypeExtension: XMLSchemaSimpleTypeDerivation {
  XMLSchemaSet<XMLSchemaAttribute>  attributes();
}
```

In a complex type derivation the base type is complex, hence the result type of the method base should be overridden covariantly, but we are forced to use the previously described C# technique. In a complex type derivation additional attributes may be added and the new particle structure is specified:

```
interface XMLSchemaComplexTypeDerivation: XMLSchemaTypeDerivation {
  new XMLSchemaComplexType  base();
  XMLSchemaSet<XMLSchemaAttribute> attributes();
  XMLSchemaParticle  particle();
}
```

A complex type extension amounts to extending the particle structure of the base type. The new particle structure is a sequence group, the first component of which is the base particle, and the rest are particles that are appended.

```
interface XMLSchemaComplexTypeExtension: XMLSchemaComplexTypeDerivation {
 new XMLSchemaSequenceGroup particle();
}
```

In a complex type restriction changes may be made to the attributes, and the particle structure of the base is restricted by restricting the ranges of occurrences or omitting optional elements. The structural representation below is augmented by assertions as explained in section 6.

```
interface XMLSchemaComplexTypeRestriction: XMLSchemaComplexTypeDerivation {
//restricted attributes and particle structure
}
```

XSD allows structural specification of typical database integrity constraints such as uniqueness, keys and referential integrity. In XSD these constraints are called identity constraints, modeled by an XSD schema interface XMLSchemaIdentityConstraint given below. We use assertions to specify these constraints as explained in section 6. An identity constraint has a name, a selector that specifies the XML structure for which the constraint holds, and a sequence of fields whose values will have the desired property. The selector is specified by a simple XPath expression. These expressions will be instances of the type XMLPath. A referential integrity constraint requires an additional reference to a key which is given by the key name.

```
interface XMLSchemaIdentityConstraint: XMLSchemaObject {
   XMLString name();
   XMLSchemaSequence<XPath> fields();
   XMLPath selector();
}
interface XMLSchemaKeyRef: XMLSchemaIdentityConstraint {
   XMLString referTo();
}
```

6   Object-Oriented Constraints

The fundamental requirement for automatic static verification dictates some limitations on expressiveness of Spec# constraints. In this section we show that these limitations fit precisely the XML Schema constraints. On the other hand, Spec# allows specification of application oriented constraints that are not XML Schema constraints.

A frequent problem in object-oriented programs is an attempt to dereference a null reference. If this happens in a database transaction, the transaction may fail at run-time with nontrivial consequences. The Spec# type system allows specification of non-null object types. Static checking will indicate situations in which an attempt is made to access an object via a possibly null reference. Examples presented in this paper include a non-null reference to a job to be inserted in the sequence of jobs, and a non-null reference to a job identifier when updating or deleting a job. Further examples of the non-null constraint are references to a sequence of jobs and a sequence of job offers which are required to be non-null.

A method in Spec# is in general equipped with a precondition expressed by the `requires` clause, and a postcondition specified by the `ensures` clause. A postcondition in general refers both to the object state before method execution, denoted by the keyword `old`, and the object state after method execution. A class is in general equipped with an invariant which specifies valid object states outside of method execution. These assertions allow usage of universal and existential quantifiers as in first-order predicate calculus, as well as combinators typical for database languages such as min, max, sum, count, avg etc.

Spec# constraints limit universal and existential quantification to variables ranging over finite integer intervals. Such intervals are in XML Schema determined by `minOccurs` and `maxOccurs` and thus it is possible to specify Spec# constraints that apply to finite sequences of particles. This is precisely what is needed for specification of keys and referential integrity in XML Schema. The limitation that quantifiers are restricted to integer variables ranging over finite intervals was a design decision to sacrifice expressiveness in order to allow automatic static verification. As explained above, this limitation is no problem in the application considered in this paper.

To make the job of the verifier possible, Spec# requires specification of the frame conditions for methods that change the object state, such as database updates. This is done by the `modifies` clause, which specifies those objects and their components that are subject to change. The frame assumption is that these are the only objects that will be affected by the change, and the other objects remain the same. An attempt to assign to the latter objects will be a static error.

One of the features that makes Spec# suitable for database applications is explicit support for the aggregation abstraction. A complex object is represented by its root object called the owner along with references to the immediate components of the owner specified as [Rep] fields. This way a complex object is defined as a logical unit that includes all of its components, direct and indirect. Object invariants may now be specified in such a way that they refer both to the owner object and to its components defined by the [Rep] fields.

Yet another feature that makes Spec# suitable for database transactions is an explicit mechanism for allowing methods to temporarily violate object invariants. This typically happens in database transactions where the integrity constraints are violated during transaction execution and then the constraints are reinstated when the transaction is completed and enforced at commit time. For example, the structure of the job deletion transaction presented in this paper has the following form:

```
expose(schema){
delete job;
delete offers that refer to the deleted job;}
```

After the first action of job deletion the referential integrity constrains are temporarily violated to be reinstated after the second action of deletion of the related job offers. The purpose of the expose block is precisely to indicate that the schema object invariant may be violated in this block. Otherwise, the verifier will indicate violation of the schema invariant. In the expose block the object is assumed to be in a mutable state and hence violation of the object invariant is allowed. Outside of the expose block every assignment that violates the invariant will be a static error.

The code in this section and in section 7 is for presentation purposes and differs somewhat from the actual Spec# code. A very simple example is the range-of-salary constraint given in the class JobType.

```
public class JobType : XMLSequenceGroupClass {
//constructor
//definition of properties JobID, jobTitle and salary
 invariant salary >0 && salary < 500000;
}
```

The salary range constraint can be strengthened in the subtype WellPaidJobType in accordance with the rules of behavioral subtyping. Type derivation by restriction in XML Schema is based on a similar idea.

```
public class WellPaidJobType: JobType {// constructor
 invariant salary >= 100000;
}
```

Two typical XML Schema constraints are given in the class JobSequence. One of them specifies the range of occurrences of jobs in the sequence of jobs in terms of properties minOccurs and maxOccurs. XMLSequenceGroupClass implements properties minOccurs and maxOccurs. Here we override them so that they will denote the actual range of occurrences. The other constraint specifies that the property JobID is a key for the sequence of jobs. This key constraint requires universal quantification that is expressible in Spec# with the already explained limitations, but then it is statically verifiable. A distinctive feature of the Spec# type system and its verification technology is non-null types illustrated by the type List<JobType>!.

```
public class JobSequence : XMLSequenceGroupClass {
// constructor
 [Rep]   [ElementsRep] List<JobType>! jobs;
 [Pure] public List<JobType> jobList{get{ return jobs;}}
 [Pure] public new int minOccurs{get{return 0;}}
 [Pure] public new int maxOccurs{get{return (jobs.Count-1);}}
```

```
invariant minOccurs >=0 && maxOccurs < 1000 ;
invariant jobs.Count <= maxOccurs - minOccurs +1;
invariant forall {int i in (minOccurs..maxOccurs),
               int j in (minOccurs..maxOccurs);
           jobs[i].JobID.Equals(jobs[j].JobID) ==>
                               jobs[i].Equals(jobs[j])};
}
```

In the above specification we see two cases of the aggregation abstraction as supported by the Spec# ownership model. The attribute [Rep] indicates that a list of jobs is a representation of a job sequence so that an object of type JobSequence is the owner of this list. Moreover, the attribute [ElementsRep] indicates that list elements are components of the list object which is their owner. These elements are then peers according to the Spec# ownership model. This has implications on invariants that can now be defined to apply to entire complex objects, i.e., including their components determined by the [Rep] and [ElementsRep] fields. These are called ownership-based invariants.

A sequence of job offers has a similar representation with some additional complexity. It has the range constraint for the number of offers and a key constraint on JobID. In addition, it has a referential integrity constraint which specifies that a job offer must refer to an existing job in the given list of jobs out of which the offers are constructed. This referential integrity constraint requires both universal and existential quantification expressible and verifiable as shown in the last invariant of the class OfferSequence.

```
public class OfferType : XMLSequenceGroupClass {
// constructor
// definition of properties  JobID, candidate name and SSN
}
public class OfferSequence:  XMLSequenceGroupClass {
  // constructor
 [Rep][ElementsRep] List<OfferType>! offers;
 [Rep] JobSequence! jobseq;
 [Pure] public List<OfferType>! joffers {get{return offers;}}
 [Pure] public new int minOccurs{get{return 0;}}
 [Pure] public new int maxOccurs{get{return (offers.Count-1);}}

invariant minOccurs >=0 && maxOccurs < 100;
invariant offers.Count <= maxOccurs - minOccurs +1;
invariant forall {int i in (minOccurs..maxOccurs),
               int j in (minOccurs.. maxOccurs);
         offers[i].JobID.Equals(offers[j].JobID) ==>
                               offers[i].Equals(offers[j])};
invariant forall {int i in (minOccurs..maxOccurs);
         exists {int j in (jobseq.minOccurs..jobseq.maxOccurs);
             jobseq.jobList[j].JobID.Equals(offers[i].JobID)}};
}
```

Representation of type derivation by restriction as defined in XML Schema is illustrated in the class ShortListed in which the range of occurrences of job offers is narrowed with respect to the range of occurrences in the base class OfferSequence. This pattern fits precisely the discipline of behavioral subtyping as implemented in Spec#.

```
public class ShortListed: OfferSequence {
invariant minOccurs >=1 && maxOccurs <= 10;
}
```

7   Transaction Verification

Our final contribution is integration of the technologies presented in this paper into an implemented model of automatic static verification of object-oriented transactions with respect to the object-oriented representation of XSD schemas equipped with constraints. To our knowledge this is the first time this was possible for a full-fledged mainstream object-oriented language and object-oriented schemas and transactions extended with very general constraints. The components of this model are more sophisticated features of the type system such as bounded parametric polymorphism available and statically verifiable in C#, representation of XML Schema constraints, pre and post conditions for transactions in Spec#, and their automatic static verification.

In our approach, the class Transaction is bounded parametric, where the bound type is the type of schema to which a specific transaction type is bound.

```
interface Schema {//. . .}
class Transaction<T> where T: Schema {
Transaction(T! schema) {. . .}
T schema(){get{return schema;}}
//. . .
}
```

With the examples developed in the previous section the simplest way of specifying a schema of job offers is an aggregation of a sequence of jobs and a sequence of job offers.

```
class JobSchema: Schema {
 [Rep] public JobSequence! jobsSeq;
 [Rep] public OfferSequence! offerSeq;
 JobSchema(JobSequence! jobs, OfferSequence! offers){
        this.jobsSeq = jobs;
        this.offerSeq= offers;  }
}
class JobTransaction: Transaction<JobSchema> {//...}
```

A transaction that creates (inserts) a new job and maintains the schema integrity constraints is JobInsert given below. The precondition of this transaction includes a constraint on the admissible range of salaries and a condition which guarantees that the insertion would not violate the key constraint. Yet another constraint requires that the argument is in fact a non-null pointer to a job object. The postcondition guarantees that the insertion is actually performed.

```
class JobInsert: JobTransaction {
// constructor
 void addJob(JobType! job)
 modifies schema.jobsSeq;
 requires forall {int i in
  (schema.jobsSeq.minOccurs..schema.jobsSeq.maxOccurs)
   !schema.jobsSeq.jobList[i].JobID.Equals(job.JobID)};
 requires job.salary >0 && job.salary < 500000;
 ensures exists {int j in
  (schema.jobsSeq.minOccurs..schema.jobsSeq.maxOccurs);
    (schema.jobsSeq.jobList[j].Equals(job))};
 {expose(schema.jobsSeq)
   {schema.jobsSeq.jobList.Add(job);}}
}
```

A salary update transaction given below takes a non-null `jobId` pointer and requires that this `jobId` actually appears in the list of jobs. The postcondition guarantees the salary update is performed correctly.

```
class SalaryUpdate: JobTransaction {
// constructor
 void updateSalaries(string! jobId)
 modifies schema.jobsSeq;
 requires exists {int j in
  (schema.jobsSeq.minOccurs..schema.jobsSeq.maxOccurs);
    schema.jobsSeq.jobList[j].JobID.Equals(jobId)};
 ensures forall {int j in
  (schema.jobsSeq.minOccurs..schema.jobsSeq.maxOccurs);
    schema.jobsSeq.jobList[j].JobID.Equals(jobId)==>
       schema.jobsSeq.jobList[j].salary >= 100000 };
 {expose(schema.jobsSeq)
  {foreach (JobType! job in schema.jobsSeq.jobList){
      if ((job.JobID.Equals(jobId)) && (job.salary < 100000))
        {job.salary= 100000;}}}  }
}
```

The most complex example of a transaction is `JobDeletion`. This transaction deletes a job with an existing given `jobId`. This requirement is expressed in the precondition. There are two postconditions. The first one guarantees that there is no job with the given `jobId` in the list of jobs, i.e., the job has been deleted. The other postcondition guarantees that the referential integrity is maintained, i.e., this `jobId` does not appear in the list of offers either.

```
class JobDeletion: JobTransaction {
// constructor
 void deleteJobs(string! jobId)
 modifies schema.jobsSeq, schema.offerSeq;
 requires exists {int j in
  (schema.jobsSeq.minOccurs..schema.jobsSeq.maxOccurs);
              schema.jobsSeq.jobList[j].JobID.Equals(jobId)};
 ensures forall {int j in
```

```
  (schema.jobsSeq.minOccurs..schema.jobsSeq.maxOccurs);
            !schema.jobsSeq.jobList[j].JobID.Equals(jobId)};
ensures forall {int j in
(schema.offerSeq.minOccurs..schema.offerSeq.maxOccurs);
   !(schema.jobsSeq.jobList[j].JobID.Equals(jobId))};

{expose(schema)
  {foreach (JobType! job in schemaJob.jobsSeq.jobList)
  if (job.JobID.Equals(jobId))
     schema.jobsSeq.jobList.Remove(job);
  foreach (OfferType! job in schema.offerSeq.joffers)
  if (job.JobID.Equals(jobId))
       schema.offerSeq.joffers.Remove(job); }}
}
```

As of this writing, the Spec# implementation is a prototype with problems that one can naturally expect from software that is still not a product. But even where static verification does not succeed, the Spec# compiler generates code that enforces the constraints at run-time, which is the prevailing technique in current object-oriented assertion languages such as JML and Eiffel.

8   Conclusions

As a rule, object-oriented application programmers have very limited understanding of what XML Schema is all about. The reason is the complexity of XSD and its mismatch with object-oriented languages. Our initial contribution is the design of an object-oriented interface to the structural core of XSD which has not been available so far. The presented collection of interfaces constitutes a library which database designers, object-oriented application programmers, and users writing queries can understand and use in developing their applications that manage data that conforms to XSD.

More importantly, we argue that the only way to resolve major issues in the object-oriented/XML mismatch is to make use of an object-oriented assertion language that allows specification of constraints-based features of XML Schema. This approach has an additional major advantage: it allows specification of more general constraints in object-oriented schemas that reflect constraints in the application environment and are not expressible in common database technologies.

Most importantly, the assertion language that we used in this paper comes with automatic verification not available in other object-oriented assertion languages. This means that for the first time we can specify object-oriented schemas and transactions equipped with general constraints and carry out static automatic verification of transactions with respect to the specified constraints. The implications on data integrity, efficiency and reliability of transactions are obvious and non-trivial.

However, the presented technology has its limitations. Spec# is a promising development, but at the moment it is an industrial prototype and not a product. Its current limitations in expressiveness dictated by the requirement for automatic static verification did not present a problem in our quite complex application. But the error messages and the current tutorial [13] need improvement, which often makes it hard to drive programs through verification. The ownership model is complex, which many users may find hard to fully understand and apply correctly. All of this implies that Spec# run-time checks for assertions that have not been statically verified is at the moment an important feature of this technology. But the technology itself is clearly still under development.

References

1. Alagić, S., Royer, M., Briggs, D.: Verification technology for object-oriented/XML transactions. In: Norrie, M.C., Grossniklaus, M. (eds.) ICOODB 2009. LNCS, vol. 5936, pp. 23–40. Springer, Heidelberg (2010)
2. Alagić, S., Bernstein, P.: An object-oriented core for XML Schema, Microsoft Research Technical Report MSR-TR-2008-182 (December 2008),
   http://research.microsoft.com/apps/pubs/default.aspx?id=76533
3. Alagić, S., Bernstein, P.: Mapping XSD to OO schemas. In: Norrie, M.C., Grossniklaus, M. (eds.) ICOODB 2009. LNCS, vol. 5936, pp. 149–166. Springer, Heidelberg (2010)
4. Alagić, S., Royer, M., Briggs, D.: Verification theories for XML Schema. In: Bell, D.A., Hong, J. (eds.) BNCOD 2006. LNCS, vol. 4042, pp. 262–265. Springer, Heidelberg (2006)
5. Alagić, S., Logan, J.: Consistency of Java transactions. In: Lausen, G., Suciu, D. (eds.) DBPL 2003. LNCS, vol. 2921, pp. 71–89. Springer, Heidelberg (2004)
6. Benzanken, V., Schaefer, X.: Static integrity constraint management in object-oriented database programming languages via predicate transformers. In: Aksit, M., Matsuoka, S. (eds.) ECOOP 1997. LNCS, vol. 1241, pp. 60–84. Springer, Heidelberg (1997)
7. Data Contracts, http://msdn.microsoft.com/en-us/library/ms733127.aspx
8. Document Object Model (DOM), http://www.w3.org/TR/REC-DOM-Level-1/
9. Lammel, R., Meijer, E.: Revealing the X/O impedance mismatch, Datatype-Generic Programming. In: Backhouse, R., Gibbons, J., Hinze, R., Jeuring, J. (eds.) SSDGP 2006. LNCS, vol. 4719, pp. 285–367. Springer, Heidelberg (2007)
10. Language Integrated Query, Microsoft Corporation,
    http://msdn.microsoft.com/en-us/vbasic/aa904594.aspx
11. Microsoft Corp., LINQ to XML,
    http://msdn.microsoft.com/en-us/library/bb387098.aspx
12. Microsoft Corp., LINQ to XSD Alpha 0.2 (2008),
    http://blogs.msdn.com/xmlteam/archive/2006/11/27/
    typed-xml-programmer-welcome-to-LINQ.aspx
13. Microsoft Corp., Spec#, http://research.microsoft.com/specsharp/
14. Microsoft Corp., XML Schema etc. Object Model (SOM) (vs.71).aspx,
    http://msdn2.microsoft.com/en-us/library/bs8hh90b
15. Shariar, Md.S., Liu, J.: Checking satisfaction of XML referential integrity constraints. In: Liu, J., Wu, J., Yao, Y., Nishida, T. (eds.) AMT 2009. LNCS, vol. 5820, pp. 148–159. Springer, Heidelberg (2009)
16. Sheard, T., Stemple, D.: Automatic verification of database transaction safety. ACM Transactions on Database Systems 14, 322–368 (1989)

17. Spelt, D., Even, S.: A theorem prover-based analysis tool for object-oriented databases. In: Cleaveland, W.R. (ed.) TACAS 1999. LNCS, vol. 1579, pp. 375–389. Springer, Heidelberg (1999)
18. W3C: XML Schema 1.1, `http://www.w3.org/XML/Schema`
19. XML Data Binder, `http://www.liquid-technologies.com/XmlStudio/Xml-Data-Binder.aspx`
20. XMLBeans, `http://xmlbeans.apache.org`

Solving ORM by MAGIC:
MApping GeneratIon and Composition

David Kensche, Christoph Quix, Xiang Li, and Sandra Geisler

RWTH Aachen University, Informatik 5 (Information Systems), 52056 Aachen, Germany
{kensche,quix,lixiang,geisler}@dbis.rwth-aachen.de

Abstract. Object-relational mapping (ORM) technologies have been proposed as a solution for the impedance mismatch problem between object-oriented applications and relational databases. Existing approaches use special-purpose mapping languages or are tightly integrated with the programming language. In this paper, we present *MAGIC*, an approach using bidirectional query and update views, based on a generic metamodel and a generic mapping language. The mapping language is based on second-order tuple-generating dependencies and allows arbitrary restructuring between the application model and the database schema. Due to the genericity of our approach, the core part including mapping generation and mapping composition is independent of the modeling languages being employed. We show the formal basis of *MAGIC* and how queries including aggregation can be defined using an easy to use query API. The scalability of our approach is shown in the evaluation using the TPC benchmark.

1 Introduction

A common design pattern for current information systems is an architecture in which a set of object-oriented classes (in the following called the *application model*), representing business objects, is stored in a relational database that conforms to a relational schema. The heterogeneous models are tailored for different requirements. The relational schema is usually specified with strong efficiency requirements in mind, whereas the object-oriented application model aims at abstraction, extensibility and maintainability of the application. This usage of different modeling languages gives rise to some problems summarized as the *object-relational impedance mismatch* [6,15].

The application model, implemented in Java classes, contains different kinds of model elements such as simple properties, multi-valued properties, or associations between classes. However, some of these model elements are not available in the relational model which can only express flat first normal-form relations with simple single-valued attributes. Consequently, the relational representation must somehow mimic such constructs. For instance, an inheritance relationship may in one case be mapped to a boolean attribute, or in another case to some enumeration of disjoint subtypes. Application developers want to query objects from the underlying database by specifying queries over the object model, but the database requires SQL queries to be posed against its relational schema. Properties of objects are manipulated by the application. These updates must also be propagated to the relational datastore. Thus, using a relational database to store objects in the application model requires posing queries against the relational

A. Dearle and R.V. Zicari (Eds.): ICOODB 2010, LNCS 6348, pp. 118–132, 2010.

schema and *unmarshalling* the objects, i.e., transforming the fetched records into instances of classes in the application model. Storing objects requires translation from the object-oriented model to the underlying relational schema. These tasks are tedious and error-prone; studies reported that 30-40% of the total project effort are spent on implementing object-relational data access [7]. *Object-relational mapping* (ORM) tools help solving these problems by mapping between the two paradigms and providing querying and updating capabilities.

Existing ORM tools use special-purpose mapping languages (e.g., Hibernate) or are tightly integrated with the programming language (e.g., LINQ [13]), which avoids the general application of such technologies in other languages which are not based on the .NET framework. In this paper, we present our ORM framework *MAGIC* (MApping GeneratIon and Compilation), that we developed based on our generic metamodel *GeRoMe* [8] and its formal *generic schema mappings* [10].

In MAGIC, the developer performs queries and updates against the object-oriented application model by creating instances of a special Query class. The query results are instantiated as objects of the application model. Updates on these objects can be propagated to the underlying database. Details of accessing and manipulating data in the underlying database are hidden by the MAGIC interface. This hiding of the persistent storage is achieved through a pair of declarative *generic schema mappings*, the *query* and *update views*. Our generic mapping language allows arbitrary restructuring of the data between the application model and the database schema, as it is based on second-order tuple-generating dependencies supporting grouping and nesting of data [10]. The object-oriented Query is represented by a *query mapping* in this generic mapping language that is rewritten into a query against the relational model by composing it with the query view. The resulting generic query can be translated into a SQL query for the relational database. As queries and mappings are represented in our generic mapping language, independently of a specific query and modeling language, they can be translated also into another concrete query language such as XQuery. Thus, our approach is not limited to relational databases as persistent storage, also other sources such as XML databases can be accessed.

The contribution of this paper is a generic mapping framework to solve the impedance mismatch between object-oriented applications and the persistent data storage. The expressivity of the mapping language allows complex restructuring between the application model and the database schema, and thus enabling more complex mappings than just 1:1 mappings between classes and tables as in most other approaches. Due to the genericity of our modeling and mapping languages, other mapping problems such as object-XML mappings can be solved, too.

The paper is structured as follows. In section 2, we introduce the simple generic metamodel and the generic mapping language, which are the basis of MAGIC. The main functionality of MAGIC, the ability to query stored data using an object-oriented query API is described in section 3. This section also explains query views and query mappings and their composition. Section 4 presents how updates can be performed using MAGIC. The scalability of MAGIC is shown in the evaluation in section 5. Related work is discussed in section 6, before we conclude the paper in section 7.

2   Background

In this section, we will briefly present a simple generic model representation which is a subset of our more comprehensive generic metamodel *GeRoMe* [8]. The simple meta-model is sufficient for the purpose of this paper. We sketch the formal semantics for our simple metamodel representation based on the semantics of *GeRoMe* [8]. The representation of generic schema mappings taken from [10] is based on this representation of schema instances. Due to space constraints, we focus on showing the key features using a running example; the formal definitions are elaborated in [8,10].

2.1   A Simple Generic Model Representation

We identified a set of modeling constructs which are common in most data model-ing languages: complex types with attributes and associations with association ends. A model (or schema) consists of *complex types*, such as XML Schema complex types or Java classes. A complex type may have *attributes*; the type of an attribute is either a simple *domain* (e.g. integer, float) or a complex type. *Associations* are relationships between complex types, or between a complex type and a domain. The degree of an association is at least two. Each participating type is connected to the association via an *association end*. There is a specialized association end that denotes composition re-lationships (a *composition end*), i.e. an instance of the component type cannot exist without being a component of a single parent type instance.

There are two types of constraints which both can be attached to attributes and asso-ciation ends: cardinality constraints and uniqueness (key) constraints.

Fig. 1 gives examples how different modeling languages can be mapped to such a generic model representation. In the examples we abstain from showing simple types of attributes, since they are not important in our context. For visualization we adopt an informal ER-like notation showing complex types as boxes, attributes as ellipses, associations as diamonds and association ends as arrows pointing to the participating types. We assume min and max cardinality of all attributes to be 1. Keys are underlined.

Relational Schemas: Relations are modeled as complex types with attributes. In 1NF only domains are allowed as attribute types, whereas in the nested relational model at-tributes may have complex types themselves. All attributes have a maximum cardinality of 1, and a minimum cardinality of 0 or 1.

Object-oriented models: A class is modeled as a complex type. Properties with sim-ple types are represented as attributes with domains. References to other classes are represented as association ends with the property name as role name.

The representation of the simple example schemas is straightforward, but also more complex models (such as nested relational models or XML schemas) can be repre-sented in this generic representation. For more complex examples, we refer the inter-ested reader to [8] and [10].

2.2   Instance Semantics

Schema mappings define relationships between schema instances. Therefore, a prereq-uisite for the definition of mappings is the characterization of instances for generic

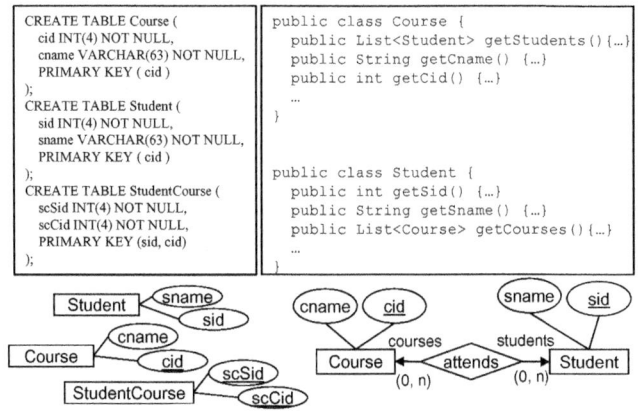

Fig. 1. Generic schema representations

1	$inst(\#1, \text{Student}) \wedge av(\#1, \text{sname}, \text{'John'}) \wedge av(\#1, \text{sid}, 123) \wedge$
2	$inst(\#2, \text{attends}) \wedge part(\#2, \text{students}, \#1) \wedge part(\#2, \text{courses}, \#3) \wedge$
3	$inst(\#3, \text{Course}) \wedge av(\#3, \text{cname}, \text{'Databases'}) \wedge av(\#3, \text{cid}, 456)$

Fig. 2. Example generic instance for the Java model in fig. 1

models, i.e., a semantics for the modeling language described. An instance of a model is a set of abstract objects with specific properties and relationships to other objects. We describe the objects, their properties and relationships by a set of logical facts.

Fig. 2 shows an example instance of the Java model in fig. 1. Each "feature" of an instance object is represented by a separate fact. The *abstract* IDs (e.g., $\#1$, $\#2$, and $\#3$) connect these features so that the complete object can be reconstructed.

The first line defines an instance of the Student class with two attributes, e.g., $inst(\#1, \text{Student})$ declares $\#1$ to be an instance of Student, and the av predicates define values for the attributes sname and sid. The av predicates are a shortcut for a combination of *attr* and *value* predicates: $av(id_1, a, v) \Leftrightarrow \exists id_2 attr(id_1, a, id_2) \wedge value(id_2, v)$. Thus, also values are represented by *abstract* objects, but we will usually omit these value objects and use the av predicates for simplification. Line 2 defines an instance of the association with two participators, the Student object defined in line 1 and the Course object in line 3. Participation in an association is represented by the *part* predicates. Please note that all $inst$ predicates in this example are implied by av and $part$ predicates, because the underlying model defines the complex types that own the attributes, the associations that own association ends, and participating types of association ends, respectively. Therefore, we will often omit the $inst$ predicates in the following examples.

The model representation is not only able to define flat structures like tables, but also hierarchical structures, e.g., element hierarchies in XML schemas. In the following subsection, this representation is applied to SO tgds which results in an expressive, generic, composable, and executable mapping language.

$$\exists s, c, sc$$
$$(\forall o_0, SI, SN$$
$$inst(o_0, Student_D) \wedge av(o_0, sid_D, SI) \wedge av(o_0, sname_D, SN) \rightarrow$$
$$\quad inst(s(SI), Student_A) \wedge av(s(SI), sid_A, SI) \wedge av(s(SI), sname_A, SN)) \wedge$$
$$(\forall o_0, CI, CN$$
$$inst(o_0, Course_D) \wedge av(o_0, cid_D, CI) \wedge av(o_0, cname_D, CN) \rightarrow$$
$$\quad inst(c(CI), Course_A) \wedge av(c(CI), cid_A, CI) \wedge av(c(CI), cname_A, CN)) \wedge$$
$$(\forall o_0, SI, CI$$
$$inst(o_0, StudentCourse_D) \wedge av(o_0, scSid_D, SI) \wedge av(o_0, scCid_D, CI) \rightarrow$$
$$\quad inst(s(SI), Student_A) \wedge inst(c(CI), Course_A) \wedge$$
$$\quad inst(sc(SI, CI), Attends_A) \wedge$$
$$\quad part(sc(SI, CI), students_A, s(SI)) \wedge part(sc(SI, CI), courses_A, c(CI))))$$

Fig. 3. Query views between relational and object-oriented schemas

2.3 Generic Schema Mappings

Having defined a generic metamodel and its formalization on the instance level it is possible to define a mapping language between two such models. This yields a generic mapping language that is agnostic about the native modeling languages. We adopt here the mapping language defined in [10] which, while being generic, still fulfills many important requirements [2] such as rich expressivity and executability. This allows generic solutions of model- and mapping-intensive problems. The mappings are second-order tuple-generating dependencies (SO tgds, [5]) with a limited set of predicates (e.g., those used in the example above of fig. 2). A mapping is an expression of the form $\exists \mathbf{f}((\forall \mathbf{x_1}(\varphi_1 \rightarrow \psi_1)) \wedge \ldots \wedge (\forall \mathbf{x_n}(\varphi_n \rightarrow \psi_n)))$ in which $\mathbf{f}$ is a set of function symbols, each $\mathbf{x_i}$ is a set of variables, and φ_i and ψ_i are conjunctions of atomic predicates. Predicates in φ_i refer to a source model $\mathbf{S}$ and might include equality predicates, predicates in ψ_i refer to a target model $\mathbf{T}$. Thus, our mappings are source-to-target.

Fig. 3 shows as an example mapping the *query views* for our small university example. In the relational database, we have three tables, *Student*, *Course*, and a table representing the connection between students and courses. Consequently, we have three implications, one for each of the classes and the last one that maps only the association. To distinguish the model elements in the database schema and the application model, we indexed the elements with D and A, respectively.

To describe the structure of instances on the target side, we have to provide a grouping functionality. This is enabled by using *abstract functions*, which are interpreted as Skolem functions, i.e., they are only instantiated as ground terms that uniquely identify instances of model elements on the target side. Consequently, the choice of arguments for an abstract function determines the grouping behaviour defined by the mapping. In the example, s, c, and sc are abstract functions to identify the objects on the target side. Abstract functions can also be used to merge data on the target side, as the same function symbol can be used in different implications; then, each implication $\varphi_i \rightarrow \psi_i$ gives only a partial description of the target object.

Our mapping language allows also concrete functions (e.g., value transformations or string concatenation) which are evaluated during mapping execution. An important feature of our mapping language is that they are closed under composition, i.e., when we compose two generic schema mappings, we will get a valid generic schema mapping as

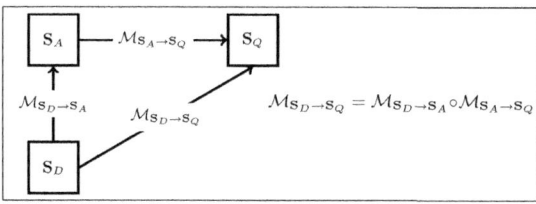

Fig. 4. Querying in MAGIC

result. This property of the mappings is the formal foundation for our object-relational mapping approach, as mapping composition will be used to perform query rewriting.

The corresponding *update views*, i.e., mappings from the application model to the database, are in the example easy to define by just reversing the implications in the query views. Please note, that inverting schema mappings might be more complex [4].

3   Querying in MAGIC

Fig. 4 depicts the process of querying a database with MAGIC. It shows the relational schema S_D, the application model S_A, and a *query model* S_Q. Moreover, the figure depicts three different mappings, the *query views* $\mathcal{M}_{S_D \to S_A}$, e.g., the mapping from fig. 3, the *query mapping* $\mathcal{M}_{S_A \to S_Q}$, and the *query composition mapping* $\mathcal{M}_{S_D \to S_Q}$.

The query model describes the structure of the query results. This model is built according to the definition of the query mapping $\mathcal{M}_{S_A \to S_Q}$ for each query. The query model contains elements from the application model as the results of the query are usually instances of classes in the application model. The consequent of the query mapping describes the objects that must be generated from the data returned by the query. Thus, the query model describes the part of the application model required to be instantiated.

The query view $\mathcal{M}_{S_D \to S_A}$ and also its inverse, the update view $\mathcal{M}_{S_A \to S_D}$ have to be provided as input to MAGIC. We provide a comfortable graphical mapping editor as part of our holistic model management system *GeRoMeSuite* [9].

To translate the object-oriented query represented by the *query mapping* $\mathcal{M}_{S_A \to S_Q}$ to a new query against the underlying database, $\mathcal{M}_{S_A \to S_Q}$ is composed with the *query view* $\mathcal{M}_{S_D \to S_A}$. The result is the *query composition mapping* $\mathcal{M}_{S_D \to S_Q}$ that maps directly from the relational database schema to the query model. Our model management system *GeRoMeSuite* provides a function to export such a mapping into a SQL query to fetch the data from the database that is required for populating instances of classes in the query model [10]. Thus, there are several steps to be taken when executing a query against the object-oriented application model:

Mapping Generation: Specify an object-oriented query against the application model by creating a Query object and invoking appropriate methods of this object.

Mapping Composition: Compose the resulting *query mapping* $\mathcal{M}_{S_A \to S_Q}$ with the known *query views* $\mathcal{M}_{S_D \to S_A}$ to yield a new mapping $\mathcal{M}_{S_D \to S_Q}$ from the relational schema to the model of the objects queried for.

```
01 Query query = new Query(queryview);
02 query.addClass("S", Student.class);
03 query.addRestriction(Restrictions.eq("S.name","Helen"));
04 query.instantiateProperty("S.courses");
05 query.addResultVariable("S");
06 List<Object[]> results = query.execute();
```

Fig. 5. A simple selection query

Mapping Compilation: Compile this new *query composition mapping* $\mathcal{M}_{\mathbf{S}_D \to \mathbf{S}_Q}$ to a native SQL query.

Execution: Execute the generated native query and construct objects conforming to the query model. To do so, populate properties of the objects using *setter* methods.

3.1 Generating the Query Mapping

We provide an application programming interface (API) to generate the query mapping in the application at runtime. Method calls to the query API (class i5.modelman.magic. Query) cause predicates to be added to the query mapping. The developer can specify various features of the query, including selection conditions, selection of associations, grouping, and aggregration functions.

Fig. 5 lists the Java code for a simple selection query including retrieval of an association. After creating the Query object, the second line adds a variable named S of class Student to the query. The third line restricts the name attribute of S to the value "Helen". When selecting associated objects from an object model stored in a database, the developer must specify which edges to be retrieved from the graph. Line 4 tells MAGIC to retrieve all Course objects associated to the students with name "Helen" via the property courses which is of type List<Student>. The developer specifies variables to be returned by the query by calling the overloaded method addResultVariable(...). Called with the name of a variable that has the type of a Java class, the method selects all generated instances. Here, the query will return instances of the Student class.

When calling the execute method of the Query class, the *query mapping* will be built. For each call to addClass(var, clazz), MAGIC will add the predicate $inst(var, clazz)$ to the antecedent of the query mapping. Thus, the variable var represents an abstract variable in terms of our generic schema mappings. When $instantiate-Property(path)$ is called, appropriate *part* predicates are added to the antecedent that query for the association. Likewise, adding restrictions causes comparison predicates to be added to the query antecedent. When the query does not contain aggregation functions, MAGIC will fetch all simple properties of abstract variables specified in the query. Corresponding av predicates will be added for these simple attributes. For the example of fig. 5, the resulting query mapping is shown in fig. 6. The antecedent queries for student and course identifiers and their names. Moreover, it queries for the association between students bound to the variable S and courses bound to the variable C.

The predicates in the consequent mimic the structure defined by the antecedent. The reason for this is that the target expresses the structure of the object model retrieved,

$$
\begin{aligned}
\exists s, c, sc \quad (\forall S, C, SC, SI, SN, CI, CN \\
av(S, SId_A, SI) \wedge av(S, SName_A, SN) \wedge av(C, CId_A, CI) \wedge av(C, CName_A, CN) \wedge \\
part(SC, Students_A, S) \wedge part(SC, Courses_A, C) \wedge equals(SN, "Helen") \rightarrow \\
av(s(SI), SId_A, SI) \wedge av(s(SI), SName_A, SN) \wedge av(c(CI), CId_A, CI) \wedge \\
av(c(CI), CName_A, CN) \wedge \\
part(sc(SI, CI), Students_A, s(SI)) \wedge part(sc(SI, CI), Courses_A, c(CI)))
\end{aligned}
$$

Fig. 6. Query mapping for the query in fig. 5

```
01 Query query = new Query(queryview);
02 query.addClass("C", Course.class);
03 Aggregate agg = AggFunctions.count("C.students");
04 query.addRestriction(Restrictions.gt(agg, 9));
05 query.addGroupProperty("C.id");
06 query.addResultVariable("C.id", "CI");
07 query.addResultVariable(agg, "CC");
08 List<Object[]> results = query.execute();
```

Fig. 7. Java code for generating an aggregation query

which is exactly the query specified by the developer. The information, which abstract functions must be used with which arguments is taken from the *query views* that were given as input to MAGIC. In the query view the term $s(SI)$ identifies an instance of the Student class where SI is the value of the property SId_A. In the query mapping in fig. 6, the variable SI is used to assign a value to the same property of the Student object. Hence, the SI of the query view is unified with SI of the query mapping to yield $s(SI)$ as the abstract function term to be used in the consequent of the *query mapping*.

3.2 Aggregation Queries

Sometimes, also in object-oriented applications, the query is required to return tuples (and not objects), e.g., if aggregation functions are applied. Fig. 7 shows how to use aggregation functions and how to add a selection condition about aggregated values to the query, resulting in a HAVING expression in SQL. First, an instance of the query class is created in line 1. In line 2 the variable C of type Course is added to the query. Internally, MAGIC will add a set of predicates to the query mapping that describe the instance and its attributes. Line 3 creates an aggregation function that counts the number of students associated to the Course object C via its students property. This implicitly requires information about associated students to be retrieved as well. Line 4 adds the restriction that we are only interested in courses that have at least ten students assigned. Records in the query are to be grouped by the course identifiers (line 5). The method call addResultVariable(agg), selects the value of the aggregation function. The query defined this way corresponds to the generic query shown in fig. 8.

If aggregation functions are used it is not required to fetch all the simple properties of instances queried for. The query mapping generated for the query in fig. 7 is shown in fig. 8. Its antecedent includes av predicates for the identifiers of courses and students

$$\exists count, g \quad (\forall S, C, SC, SI, CI$$
$$av(S, SId_A, SI) \wedge av(C, CId_A, CI) \wedge$$
$$part(SC, Students_A, S) \wedge part(SC, Courses_A, C) \wedge (count(SI) > 9) \rightarrow$$
$$\mathcal{Q}(CI, count(SI)) \wedge group(g(CI), count(SI))$$

Fig. 8. An example of a *generic query with aggregation*

and queries for the association in between using *part* predicates. The restriction on the number of course participants is given as a $>$ comparison predicate involving the aggregation function.

The consequent of the query mapping in case of aggregation will contain a query head predicate including the selected variables and appropriate *group* predicates. The *group* predicates are similar to GROUP BY statements in SQL. Its first argument is defined as an abstract function defining the desired grouping behavior. The second argument is the required aggregation function.

3.3   Compiling and Executing the Mapping

When the *query mapping* $\mathcal{M}_{S_A \rightarrow S_Q}$ has been built, it is composed with the query views $\mathcal{M}_{S_D \rightarrow S_A}$ (cf. fig. 3). In doing so, the antecedent of $\mathcal{M}_{S_A \rightarrow S_Q}$ which is a query against the generic representation of the Java application model is translated to a new query against a generic representation of the database schema. The resulting *query composition mapping* $\mathcal{M}_{S_D \rightarrow S_Q}$ is a direct mapping from the relational schema to the structure of the desired result. This new mapping is then exported to the native query language. Please note that, although our mapping language can express disjunction, currently MAGIC cannot handle disjunction. To overcome this limitation the algorithm for generating native queries has to be extended to handle multiple implications as well.

When the query is executed, MAGIC will use the retrieved values to create instances of the classes in the application model, according to the *query composition mapping*. The execute method will return a list of arrays of type Object. Each array will contain one element for each variable selected by calling addResultVariable(...) in the order of the calls to this method. Thus, for the query in fig. 5 the result will be a one-dimensional array containing instances of Student with associated Courses whereas for the aggregation query defined in fig. 7, the result will be a two-dimensional array with each component array containing a course identifier in the first component and the number of students in the second component.

4   Propagating Updates to the Database

During operation, the application creates new instances of classes in the application model, associates instances with each other, and sets properties of instances. MAGIC also allows to propagate these updates to the underlying persistent storage. MAGIC offers a set of updating methods that accept objects as arguments. When calling these

```
01 Session session = new Session();
02 try {
03    Transaction tx = session.beginTransaction();
04    session.updateProperty("Student.courses");
05    session.update(student);
06    tx.commit();
07 } catch { tx.rollback();
08 } finally { session.close(); }
```

Fig. 9. Java code for updating an object `student` in the database

methods MAGIC updates the appropriate rows in the underlying database. Unlike query-ing, updating does not require a new mapping to be generated. Instead, we use the *up-date view* $\mathcal{M}_{\mathbf{S}_A \to \mathbf{S}_D}$ from the application model to the database model directly.

The method `save(obj)` must be called if `obj` is a new object that needs to be stored persistently in the database. On the other hand, the method `update(obj)` will update the tuples corresponding to the object and its associated objects in the database. In the same way as `instantiateProperty` defines which associated ob-jects to retrieve in a query, `updateProperty(...)` specifies which objects asso-ciated to `obj` shall be persisted. The method gets as input a path to the property, e.g., `"Student.courses"`. This is to avoid that every update writes the whole object graph. Usually only certain objects in the graph must be updated. Fig. 9 lists the code for updating a modified `Student` object and its associated courses.

To execute an update against a set of objects in the application model, MAGIC first finds the implications in the *update views* that contain the properties to be updated in the antecedent. Using an appropriate *update code generator* component, the required im-plications are translated to SQL `INSERT` and `UPDATE` statements in case of relational databases. Of course, the API allows to realize updates as transactions. Having gener-ated the update statements, MAGIC assigns the values of properties to the appropriate variables in the update statements and then executes the updates against the database.

Fig. 10 depicts an example SQL update code generator algorithm. By examining the $inst$ predicates of the consequent of the update view the algorithm first collects a set T of the tables which are affected. We also keep the Skolem function $f(x)$ which is used as abstract identifier of a tuple of this table. Then, the av predicates are analyzed to identify those keys and columns which are included in the mapping. Finally, the statements are created. They set all columns of the table a which are included in the update view, i.e., the sets $\mathcal{K}_{f(x)}$ and $\mathcal{C}_{f(x)}$; if there is already a tuple with the specified key, then only the columns in $\mathcal{C}_{f(x)}$ will be updated. We assume objects are uniquely determined by primary key.

Calling the methods `update(...)`, `save(...)`, and `updateProperty(...)` ex-plicitly to persist changes to the object model in the database can be inconvenient for the programmer. A possible solution uses aspect oriented programming (AOP) [11]. In AOP, such cross-cutting concerns as calling update methods of the database can be de-fined in separate modules, so called aspects. For instance, AspectJ[1] can be used to regis-ter pointcut expressions with *setter* methods of instantiated business objects. Whenever a setter method is called, this can be logged to a cache of updated properties for each

---

[1] `http://www.eclipse.org/aspectj/`

Input: An implication $\chi = \varphi \rightarrow \psi$ taken from an update view.
Output: A set of parameterized SQL select statements.
Initialization: $\mathcal{T} = \mathcal{C} = \mathcal{K} = \emptyset$
Construct the table set $\mathcal{T}$:
    for each predicate *inst(f(x), a)* in ψ
        add *(f(x), a)* to $\mathcal{T}$
Construct column set $\mathcal{C}$ and key column set $\mathcal{K}$:
    for each predicate *av(f(x), attr, term)* in ψ
        if $\mathcal{T}$ declares *attr* to be a key component for table *t*
            add *(f(x), attr, term)* to $\mathcal{K}$.
        else
            add *(f(x), attr, term)* to $\mathcal{C}$.
Construct the update:
    for each $(f(x), a) \in \mathcal{T}$
        let $\mathcal{K}_{f(x)} = \{t | t = (f(x), attr, term) \in \mathcal{K}\} =$
            $\{(f(x), attr_1, term_1), ..., (f(x), attr_m, term_m)\}$
        let $\mathcal{C}_{f(x)} = \{t | t = (f(x), attr, term) \in \mathcal{C}\} =$
            $\{(f(x), attr_{1a}, term_{1a}), ..., (f(x), attr_{na}, term_{na})\}$
        add the following SQL statement to the result
            insert into $a(attr_1, ..., attr_m, attr_{1a}, ..., attr_{na})$
            values $(?term_1, ..., ?term_m, ?term_{1a}, ..., ?term_{na})$
            on duplicate key update set $attr_{1a} = ?term_{1a}, ..., attr_{na} = ?term_{na}$

Fig. 10. Algorithm **SQLUpdateGen** with update and insert

object without changing the code of the business classes. Upon updating the object all updated simple and complex properties can be included into the update statement. Alternatively, the update of a property of an object could directly be propagated to the database. In this case, however, all update statements could be precompiled on startup of the system. Moreover, by registering appropriate aspects with the getter methods we can realize lazy fetching of associated objects and, hence, transparent persistence.

5   Evaluation

To evaluate MAGIC we performed a set of queries against 1GB of data conforming to the schema of the TPC-H benchmark. We defined a set of queries of different complexity that were answered by MAGIC. The evaluation has been performed on an Intel Core 2 Duo processor with 2.6GHz CPU and 850MB Java heap space running Windows XP. The database resided in a MySQL 5.1 server on the same system. We defined a Java object model for the TPC database, including corresponding query and update views. The following queries have been posed against the Java object model of the TPC database. To explore how MAGIC scales we varied the size of the result sets.

Query 1: The first query is a simple select query that fetches 10 to 10,000 suppliers without any associated objects. As each `Supplier` object has 6 attributes, MAGIC sets 60 to 60,000 attribute values.

Query 2: The second query selects 1 nation and 10 to 5,000 customers in that nation. MAGIC instantiates 11 to 5,001 objects and sets 10 to 50,000 attribute values.

Query 3: This query fetches 10 to 10,000 orders together with the associated customer objects. MAGIC must generate 20 to 19,521 objects from the results returned by the database and sets 140,000 attribute values for the largest result.

```
Q1:  q.addClass("S", Supplier.class);
     q.addRestriction(Restrictions.gt("S.key", 9900));
     q.addResultVariable("S");

Q2:  q.addClass("N", Nation.class);
     q.instantiateProperty("N", "customers");
     q.addRestriction(Restrictions.eq("N.name", "GERMANY"));
     q.addRestriction(Restrictions.gt("N.customers.key", 147712));
     q.addResultVariable("N");

Q3:  q.addClass("O", Order.class);
     q.addRestriction(Restrictions.leq("O.key", 388));
     q.instantiateProperty("O", "customer");
     q.addResultVariable("O");

Q4:  q.addClass("P", Part.class);
     q.addClass("S", Supplier.class);
     q.addClass("PS", PartSupp.class);
     q.addRestriction(Restrictions.leq("P.key", 100));
     q.instantiateProperty("P", "partSupps", "PS");
     q.instantiateProperty("S", "partSupps", "PS");
     q.addResultVariable("P");
```

Fig. 11. Object-oriented queries used for evaluation

Query 4: The fourth query fetches objects from three classes connected by two associations which realizes an n-to-m relationship. In particular, it returns 10 to 10,000 Parts and their Suppliers together with the PartSupp objects that connect them. The largest result set generates 60,005 objects, sets 720,072 attribute values, and creates more than 80,000 associations between the objects.

Fig. 11 depicts the Java code for these queries. For the sake of space we omitted creation and execution of the query q. The version of instantiateProperty which takes three arguments defines PS to be the PartSupp that associates P and S.

For each of these test cases, we have measured the execution time of MAGIC with the YourKit Java Profiler 7.5. The performance results in milliseconds are depicted in fig. 12. All selection predicates were defined using primary keys because our aim was to evaluate the performance of MAGIC and not the performance of the DBMS. Moreover, to avoid a bias against the DBMS, we posed each query against the database once before measuring it in 10 subsequent runs. This allows the DBMS to cache the results. The results are depicted in the figures as average values.

The figures show the total time for each query depending on the number of results, and the fractions of time needed by mapping composition, the DBMS, and MAGIC. We also compared the performance of MAGIC with Hibernate; the figures show the total time for each query in Hibernate. The composition algorithm includes various optimizations because mapping composition has exponential time complexity in general. An alternative is the more efficient *MiniCon* algorithm for rewriting [14], which we consider for future work. The time for composing the query mapping with the query views is constant for each query because the query mapping does not depend on the number of results. Consequently, the more results are returned, the more neglectable is the time for mapping composition. For small result sets, the total time required by MAGIC is less than the total time required for Hibernate. For larger result sets, the queries generated by Hibernate are significantly faster than our queries as the Hibernate queries use INNER JOIN. MAGIC produces queries with the join condition in the WHERE clause.

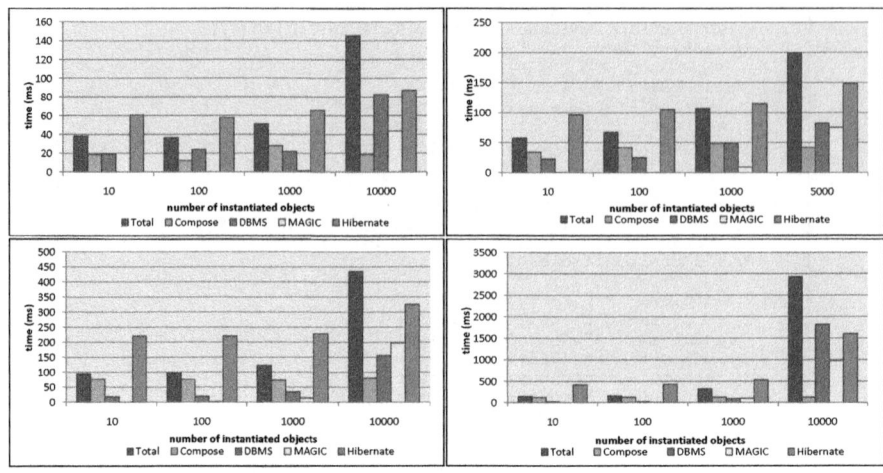

Fig. 12. Results for Queries 1 to 4

This problem holds especially for queries 1 and 4: the time required by the DBMS for the MAGIC is about the same as the total time Hibernate. This difference in the SQL query performance is probably not present in another DBMS with a different query optimizer. In summary, we can say that the performance of MAGIC is reasonable as the largest result requires only about 3 seconds to instantiate more than 60,000 objects and associating them with each other. MAGIC is scalable as the time for MAGIC increases by the same factor as the size of the result set, similar to Hibernate.

Beside the queries returning objects associated with each other, we also posed aggregation queries in our evaluation. The first aggregation query asks for the numbers of parts of each type that are offered by one particular manufacturer. This query returned 150 part types and their numbers. The second aggregation query asks for the average account balance of customers in each nation. This query returns 25 results, one for each nation. In both cases the amount of time needed by composition, and MAGIC was neglectable as compared to the time required by the DBMS to answer the queries.

6   Existing Approaches to the Impedance Mismatch

Various approaches exist for solving the object-relational impedance mismatch using different programming languages and offering different features. For the Hibernate ORM tool[2], mappings between the object model and the underlying relational database are specified as annotations of Java classes and methods or as XML files. Hibernate can perform both, queries and updates, using the information in such mapping files. Developers pose queries either in SQL against the relational schema (which does not solve the impedance mismatch) or using the HQL query language. HQL is a proprietary query language which allows specification of queries in a syntax similar to SQL. However,

---

[2] http://www.hibernate.org/

HQL is not parsed by the Java compiler. Therefore, the compiler cannot report errors in the query. Additionally, like in MAGIC query objects can be built in Java code. This approach is also more convenient for dynamic creation of query objects at runtime.

The ADO.NET Entity Framework [13,3,1] is another solution approach to the object-relational impedance mismatch. It has recently been extended for object-oriented access to XML data [16]. Queries are posed against a conceptual model specified in the proprietary modeling language *Entity Data Model* (EDM). The EDM supports associations between classes and inheritance. Mappings of superclasses are not inherited by subclasses. Thus, like in MAGIC inherited features must be mapped for both the superclass and the subclass. Moreover, ADO.NET supports the LINQ (*Language INtegrated Query*) query language. LINQ is a novel query language integrated into the .NET framework at the programming language level [12]. This allows to define queries that are statically compiled and checked by the compiler of the host programming language. Integration of ADO.NET with the LINQ language hence enables compile-time checking of declarative queries against the object model. Like our approach, the ADO.NET Entity Framework internally uses bidirectional views for rewriting queries against the object model. Views are defined as equalities between queries over the entities and queries over the relational schema. These queries are compiled to query views and update views. The views satisfy the *roundtripping criterion* which means that given a set E of entities, we have $E = V_{Query}(V_{Update}(E))$. Most notably, view maintenance techniques are employed for propagating updates to the underlying database.

In MAGIC we used similar techniques to show the usefulness of *generic schema mappings* for ORM. Although the examples presented in this section seem simple, they represent already a large class of queries required by real applications. More complex queries could also be realized using our mapping language. For instance, a developer may want to retrieve all the objects representing courses, that have more than ten participants (instead of only their identifiers). This query could be expressed in our language as well by simply adding the respective av predicates to the example of figure 8. Albeit such an extension would require changes to the Java code of MAGIC, it is not a conceptual challenge for our generic mapping language.

7 Conclusion

We presented a solution for the object-relational impedance mismatch. Our prototype *MAGIC* relies on a pair of schema mappings between the object-oriented application model and an underlying relational database. Generic queries against the OO model are rewritten into queries against the relational database by means of our mapping composition algorithm. Definition of Select-Project-Join and aggregation queries is possible through an easy to use query API. *MAGIC* generates SQL queries which are executed to retrieve a result set from which objects in the application model are instantiated without creating intermediate objects. A natural extension is to replace the SQL query generation by a component for generating XQuery. In doing, so *MAGIC* could as well provide object-oriented access to XML databases. In previous work [10], we developed already a component for generating XQueries for the core generic mapping language. This component could be easily extended to cover also the more expressive queries in this work (including aggregration). Other modeling languages could be included by providing appropriate query generation components which would require no changes to the core part of *MAGIC*, i.e., the mapping composition algorithm and query API.

Please note, that a research prototype like MAGIC cannot achieve the same rich feature set as a commercial product offered by a company that is able to extend the programming language, the programming environment, and the database system, as in the case of the ADO.NET Entity Framework. Nevertheless, our prototype shows that an ORM solution based on a generic metamodel and a generic mapping language is possible and feasible. Generic approaches have been critized because of low performance and too much overhead. However, our approach does not create any intermediate objects as our mappings can be translated directed into native query languages. Thus, we achieve a similar, for small queries even better performance than Hibernate.

Acknowledgements. The work is supported by the Research Cluster on Ultra High-Speed Mobile Information and Communcation UMIC (www.umic.rwth-aachen.de).

References

1. Adya, A., Blakeley, J.A., Melnik, S., Muralidhar, S.: Anatomy of the ado.net entity framework. In: Proc. SIGMOD Beijing, China, pp. 877–888 (2007)
2. Bernstein, P.A., Halevy, A.Y., Pottinger, R.: A vision for management of complex models. SIGMOD Record 29(4), 55–63 (2000)
3. Castro, P., Melnik, S., Adya, A.: Ado.net entity framework: raising the level of abstraction in dataprogramming. In: Proc. SIGMOD, pp. 1070–1072 (2007)
4. Fagin, R.: Inverting schema mappings. ACM Transactions on Database Systems 32(4) (2007)
5. Fagin, R., Kolaitis, P.G., Popa, L., Tan, W.C.: Composing schema mappings: Second-order dependencies to the rescue. ACM Trans. Database Syst. 30(4), 994–1055 (2005)
6. Ireland, C., Bowers, D., Newton, M., Waugh, K.: A classification of object-relational impedance mismatch. In: Proc. DBKDA, pp. 36–43. IEEE, Los Alamitos (2009)
7. Keene, C.: Data services for next-generation soas. Web Services Journal 4(12) (2004)
8. Kensche, D., Quix, C., Chatti, M.A., Jarke, M.: GeRoMe: A generic role based metamodel for model management. Journal on Data Semantics VIII, 82–117 (2007)
9. Kensche, D., Quix, C., Li, X., Li, Y.: GeRoMeSuite: A system for holistic generic model management. In: Proc. VLDB, pp. 1322–1325 (2007)
10. Kensche, D., Quix, C., Li, X., Li, Y., Jarke, M.: Generic schema mappings for composition and query answering. Data Knowl. Eng. 68(7), 599–621 (2009)
11. Kiczales, G., Lamping, J., Mendhekar, A., Maeda, C., Lopes, C.V., Loingtier, J.-M., Irwin, J.: Aspect-oriented programming. In: Aksit, M., Matsuoka, S. (eds.) ECOOP 1997. LNCS, vol. 1241, pp. 220–242. Springer, Heidelberg (1997)
12. Meijer, E., Beckman, B., Bierman, G.: LINQ: Reconciling object, relations and XML in the .NET framework. In: Proc. SIGMOD, pp. 706–706 (2006)
13. Melnik, S., Adya, A., Bernstein, P.A.: Compiling mappings to bridge applications and databases. In: Proc. SIGMOD, Beijing, China, pp. 461–472 (2007)
14. Pottinger, R., Halevy, A.Y.: Minicon: A scalable algorithm for answering queries using views. VLDB Journal 10(2-3), 182–198 (2001)
15. Russell, C.: Bridging the object-relational divide. ACM Queue 6(3), 18–28 (2008)
16. Terwilliger, J.F., Bernstein, P.A., Melnik, S.: Full-fidelity flexible object-oriented xml access. Proc. VLDB Endow. 2(1), 1030–1041 (2009)

Closing Schemas in Object-Relational Databases

Manuel Torres[1], José Samos[2], and Eladio Garví[2]

[1] Universidad de Almería, Spain
mtorres@ual.es
[2] Universidad de Granada, Spain
jsamos@ugr.es, egarvi@ugr.es

Abstract. Schema closure is a property that guarantees that no schema component has external references, that is, references to components that are not
included in the schema. In the context of object-relational databases, schema
closure implies that types, tables and views do not have references to components that are not included in the schema. In order to achieve schema closure, in
this work two basic approaches known as enlargement closure and reduction
closure are proposed. Enlargement closure includes in the schema every referenced component. Reduction closure, on the other hand, is based on the trans-
formation of the components that have external references, eliminating these
references to fulfill schema closure. In this work, both closure approaches and
the algorithms to carry out the closure in each of them are described. These algorithms generate and incorporate the needed components, whether being types
or views, to reach the schema closure making easier therefore the definition of
external schemas. Finally, to illustrate the concepts proposed in this work, we
explain how to carry out schema closure in SQL:2008.

1 Introduction

Object-relational databases (ORDB) extend relational databases with object- oriented
features, allowing user-defined types (UDT), complex objects, inheritance and so on.
From UDTs, we can build *object tables* and *object views* (also known as *typed-tables*
and *typed-views*, respectively), and include them in an object-relational schema. *Subtypes* based on existing types may also be defined, giving rise to a type hierarchy.
Analogously, *subtables* of existing tables, and *subviews* of existing views may also be
defined, giving rise to a table hierarchy and a view hierarchy, respectively. These
features have been added to the SQL standard and for example, since SQL:1999
typed-tables can be defined, each containing a set of objects accessed via methods,
and typed-views. Subtypes, subtables and subviews of existing types, tables and
views can also be defined respectively [4]. This is an enormous advance in the usefulness and applicability of SQL, reducing the impedance mismatch between databases
and programming languages.

The fact that ORDBs incorporate some object-oriented features, however, give rise
to some of the common problems of object-oriented databases (OODBs), like the
well-known *schema closure* [2,10,7,11]. Schema closure is a property that guarantees
that no schema component has *external references*, that is, references to components

A. Dearle and R.V. Zicari (Eds.): ICOODB 2010, LNCS 6348, pp. 133–146, 2010.
© Springer-Verlag Berlin Heidelberg 2010

that are not included in the schema. In this paper, the closure of object-relational schemas following two approaches, named as *enlargement* and *reduction* closure are proposed.

Basically, enlargement closure is based on including recursively each component that is referenced by schema components, so that the original schema is enlarged with the referenced components. The main drawback that can show the enlargement closure is that the systematic inclusion of referenced components in non-closed schemas may lead, in certain situations, to obscure the schema that was originally specified by the schema definer.

Reduction closure is based on the assumption that the schema definer is interested only in the components that he or she has selected, and no one else. If the schema includes some components with external references, in order to guarantee the schema closure, but without adding more components, schema components that have external references are replaced with views that hide those references.

The study of schema closure in ORDBs is important because throughout the life of the database, *external schemas* or *view schemas* may be defined to provide a particular perspective of the conceptual schema. However, the defined external schemas may be not closed. We can find an example when the schema definer has not included some of the types used by a typed-table or by a typed-view of the schema. In such a case, the schema is not closed. In this paper, we will describe how schema closure can be carried out on ORDBs. The paper also describes how schema closure can be achieved on SQL:2008 [1], because the application of reduction closure in SQL is not straightforward. This is because of reduction closure may require the generation of intermediate types and the modification of the type hierarchy (that may entail the modification of the supertypes of a type), which is not directly allowed in SQL. The remainder of this paper is organized as follows. Section 2 overviews how ORDBs may include some object-oriented features, and also introduces the example that will be used along the paper. In Section 3, the concepts of enlargement and reduction are described. Section 4 describes how closure concepts can be applied in SQL:2008. In Section 5 the related work is described. Finally, we conclude the paper in Section 6.

2   Main Features of Object-Relational Databases

In ORDBs, a schema may include tables and views as in relational databases, but in addition, object-relational schemas may include some of the features of object oriented schemas, like UDTs which consist of attributes (whose types may also be UDTs) and methods. An UDT may be defined to be a *subtype* of another UDT, known as its direct supertype. A subtype inherits every attribute and method of its direct supertype. Subtypes may also have additional attributes and methods. Besides, it is possible to redefine inherited methods.

From these UDTs, *typed-tables* can be defined. Every row of a typed-table, also known as *row object*, is an instance of the UDT that is acting as the type of the table. The typed table will have a column for each UDT attribute. Methods defined in the UDT may be applied to each table row. Typed tables can be defined as *subtables* of another typed table making up a hierarchy. For example, we could define a type person-type for persons including the attributes first-name, last-name, age, address, and ss-number. From this type, we could define a typed-table person-table to store objects whose type is person-type. In order to make the example easier, types are defined without methods. Figure 1.a illustrates how the definition of the type, the table, as well as the relationship between the table and the type can be represented. This relationship has been named as *type-of* to show that the type of the table person-table is person-type.

If a subtable student-table of person-table has to be created including the attributes IQ and entry-date, the type of the subtable must be created and positioned accordingly in the type hierarchy before. In this case, a type student-type as subtype of person-type must be created, The attributes of the new type are IQ and entry-date.

From typed-tables, object views may be defined, although object views may also be defined from existing object views. Object views allow the data customization of their underlying tables or views. To define an object view, we must take into account that the view type must be defined before defining the view, and this type must be placed accordingly in the type hierarchy.

In order to illustrate how a view can be defined in ORDBs, let us suppose that we are interested in creating an object view on person-table that selects those people less than 40 years old. This view is named as YoungerThan40-view. In this example, given that the view type is the same as the type of the base table, the view type has not to be created.

Now, if we want to define an external schema from the previous schema replacing person-table with YoungerThan40-view, the schema definer must select only the type person-type and the view YoungerThan40-view as schema components. Figure 1.b illustrates such an external schema.

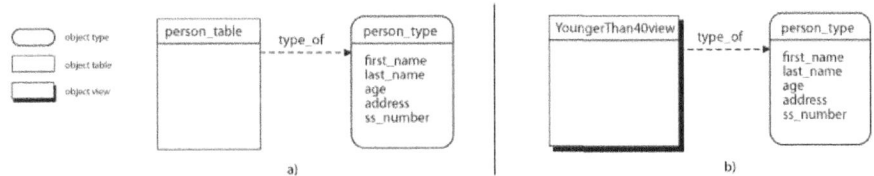

Fig. 1. a) Definition of an object type and a typed-table; b) External schema selecting those people with less than 40 years old

In order to illustrate the closure concepts that will be introduced in the next section, Figure 2 depicts the schema of a database of an insurance company that will be used along this paper. The database stores data about policies and policy covers. Each policy has one policyholder. Each policyholder is associated to his or her contact employee. Employees work in a department and they share some properties and methods with policyholders. In order to make the schema easier, *ISA* and *type-of* relationships have been depicted, but attributes and methods have not.

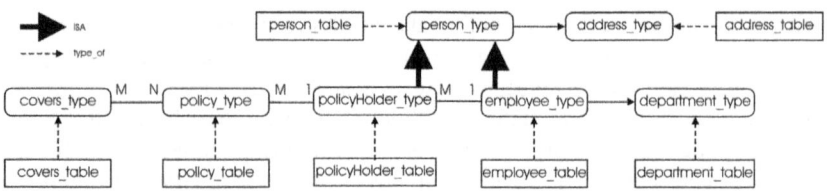

Fig. 2. Database schema for a database of an insurance company

3 Schema Closure

Schema closure is a property to guarantee that no component of a schema includes references to other components that are not included in the schema. That is, on the one hand, the types included in a schema do not have references to types that are not included in the schema; on the other hand, object tables and object views do not have references to types that are not included in the schema, and so on.

In [10], a closed schema was defined as follows: Let C be the set of components of a schema and R the set of inheritance relationships defined in it. $Uses(C_i)$ can be defined as the set of components used by the properties and operations of C_i.

Definition 1 (Closed schema). *A schema $S = (C, R)$ is closed if, and only if, $C = Uses(C_i) \cup C, \forall C_i \in C$.*

Next, we propose the closure of object-relational schemas following two approaches, named as enlargement and reduction closure.

3.1 Enlargement Closure

Enlargement closure [10] includes every component referenced by a schema component. If we apply this idea to ORDBs, each object type, table and view referenced by object types, tables and views of the schema must also be included in the schema.

In order to illustrate the operation of enlargement closure, let us suppose that from the insurance company schema of Figure 2, an external schema must be defined hiding data of employees, as well as the departments they work in. Therefore, the components selected to make up the external schema are the object tables `address-table`, `person-table`, `policyHolder-table`, `policy-table` and `covers-table`, as well as their types, as can be seen in Figure 3.a. If we observe this figure, we conclude that the schema is not closed because `policyHolder-type` has an external reference to `employee-type`. In order to close this schema following an enlargement approach, `employee-type` must also be included in the external schema. However, the inclusion of `employee-type` does not solve the problem because the schema is still not closed. Now, the schema is not closed because the just added `employee-type` has an external reference to `department-type`. Therefore, `department-type` must also be included in the schema. Once the types `employee-type` and `department-type` have been added to the external schema, the schema is now closed, as it is depicted in Figure 3.b.

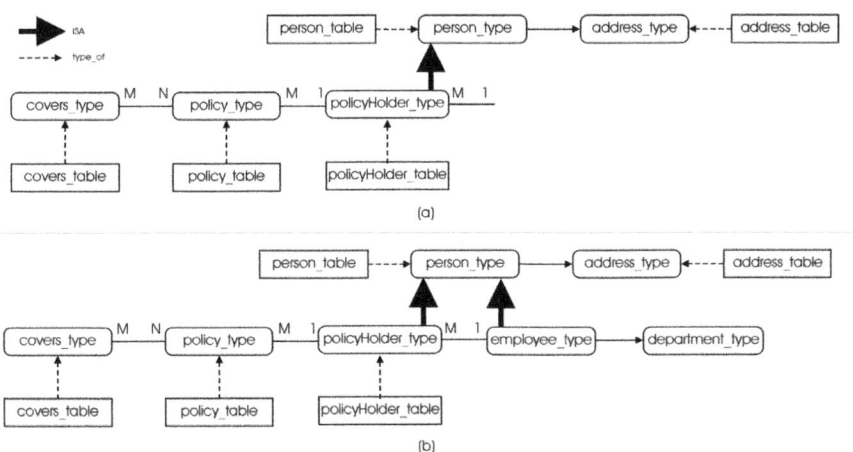

Fig. 3. a) Components selected by the schema definer; b) External schema closed following the enlargement approach

This kind of closure is based on the idea that the user wants a schema with the components that he or she has selected and, if it is necessary, other components can also be included into it in order to achieve the schema closure. Below, the enlargement closure algorithm is shown.

```
Function EnlargementClosure (S): NeededComponents
1. Temp = C = GetComponents(S); NeededComponents = ∅
2. while Temp ≠ ∅
3.    C_i = GetAndRemoveNext(Temp)
4.    if C_i ∉ C and C_i ∉ NeededComponents then
5.            NeededComponents = NeededComponents ∪ {C_i}
6.    end if
7.    for all C_k ∈ Uses(C_i)
8.            if C_k ∉ C and C_k ∉ Temp and
              C_k ∉ NeededComponents then
9.                    Temp = Temp ∪ {C_k}
10.           end if
11.   end for
12. end while
13. return NeededComponents
```

This algorithm takes as input the set of components S selected to make up the external schema. The algorithm returns in NeededComponents the components that are needed to close the schema.

The schema S only consists of the set of schema components C specified by the schema definer. This set is obtained by the function GetComponents that is not described here. This function returns the set of components that make up the schema S, that can be obtained from the data dictionary. The set of components C that make up the schema S is copied to an auxiliary variable Temp which will be used to check the schema closure.

The algorithm processes the components of Temp one by one pulling out from this set by a function GetAndRemoveNext, that is not described here. If the component that is being processed does not belong to the set of schema components or to the set of needed components (i.e. NeededComponents), it is added to the set of needed components. Checking whether NeededComponents is empty we can know if the schema is closed.

Next, the components used by the current component are added to Temp if they have not been already added; this is checked at this way: a component does not have to be added to Temp if it is a component of C (it was already included in Temp at the beginning of the algorithm), if it is a component that is already in Temp (it is a component that is not still processed) or if it belongs to NeededComponents (it is an external reference added by another component).

3.2   Reduction Closure

Enlargement closure may add some components to external schemas because all the referenced components are also included into the schema, as well as all the components referenced by them, and so on. However, in some situations the schema definer may not want to include into the external schema all the referenced components, or some of them. For example, this would mean that the schema definer selects only the components person-table and person-type, as shown in Figure 4.a, and he or she does not want to include address-type, which is referenced by person-type. However, since schema closure must be fulfilled, but no additional components are wanted, components with external references must be replaced with another component, which do no not include external references. This is the premise which reduction closure is based on, that is, to replace components that have external references in order to remove such references. Thus, schema closure is fulfilled and no components are added to the schema.

Figure 4.b illustrates the resulting external schema after applying reduction closure. Following this closure approach, person-table is replaced with an object-view person-view, which hides the external references of its base table. The definition of this object view implies the definition of a new type, person-view-type, which hides the reference to address-type that has the original person-type. The type newly defined will be the type of the object view person-view.

Fig. 4. a) Components selected by the schema definer; b) External schema closed following the reduction approach

Summarizing, reduction closure assumes that the user wants to include only the components that he or she has selected, and no one else. In order to remove the references to non-included components, new types and/or object views have to be defined to replace them. Components with external references cannot be directly modified because, if we modify them instead of defining new ones, we would produce collateral effects in other schemas where those components were also included (e.g. the conceptual schema or other external schemas).

A priori, reduction closure is only another alternative to reach schema closure. However, it can also be used as a mechanism to define external schemas because it simplifies the external schema definition process. In fact, object views and new types that hide external references are defined automatically, and existing relationships are updated automatically. However, the definition of the new components has to be scheduled carefully in order to avoid multiple modifications of the same type or the same object view, doing the needed modifications in one go. In order to obtain which are the types, tables and object views that must be replaced, we propose a set of rules to decide whether a component has to be replaced with a view component to update its references.

Obtaining the types to be modified
These rules consider the five cases that may occur, which are illustrated in Figure 5. In the figure, modified types are marked with an asterisk (*), and the dotted arrow points to affected types.

- If a type A is modified and B is an aggregated type of A, B does not need to be modified because of the modification of A. That is, aggregated types of a modified type do not need to be adapted (Figure 5.a).
- If a type A has an aggregation relationship with a type B, and B is modified, the type A must be modified. That is, if an aggregated type has to be adapted, the types that have a reference to it have to be modified after it (Figure 5.b).
- If two types A and B are related with an association, and at least, one of them has to be modified, both types have to be adapted (Figure 5.c).
- If A is a type that must be modified and B is a subtype of A, B must be modified after modifying A. That is, subtypes of a modified type must be modified after it (Figure 5.d).
- If A is a supertype of a type B to be modified, A does not need to be modified because of the modification of B. That is, supertypes of an adapted type do not need to be modified if they have not to be modified for another reason (Figure 5.e).

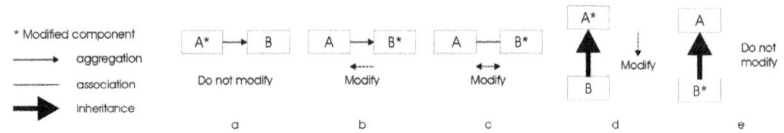

Fig. 5. Cases for propagating modifications of types in a schema

Following these indications, a list of nodes can be built. Nodes symbolize types to be modified, either because of having external references, or to propagate the modifications. The node structure is depicted in Figure 6. Each node contains the name of the type to be modified (OldTypeName) and the name of the type that is

going to replace it (NewTypeName -this name may be generated automatically adding a numerical su   x indicating the number of types defined from its base type). In
addition, the node includes the list of types included in the schema that are referenced
by the type (ReferencesTo), the list of external references that must be deleted
(ToBeDropped), and the list of references that must be updated (ToBeUpdated).
Adding the list of external references (ToBeDropped) in each node makes easier the
later definition of the type that will replace the type corresponding to that node in
order to hide the external references. The list ReferencesTo is used to check
whether a type has references to types that have been replaced with new types, and
therefore, must update its references

Obtaining the tables and object views to be modified

After applying the previous set of rules, new types have been defined to replace the
types with external references, as well as to replace those types affected by the replacement of the types with external references. Now, we must obtain the set of object
tables and object views that were defined from a type that has been replaced with a
new type. These object tables and views must also be replaced with object views
defined from them, but using the new type definitions. That is, if an object table or an
object view OV was defined originally from a type T, and this type has been replaced
with a type T', then a new object view OV' must be defined from OV. The type of
OV' is T'.

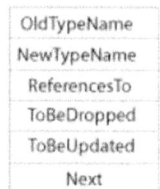

Fig. 6. Node structure corresponding to types to be modified

This set of object views may be obtained going all over the node list built following the previous set of rules. For each node in the list, if a table or object view was
defined using the type represented by the node, a new object view must be defined.
The name of the type of this new object view may be found in the NewTypeName of
the node.

Figure 7 illustrates how to obtain the object views that replace the object tables and
views whose types have been replaced with new types to reach reduction closure.

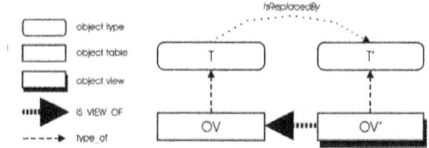

Fig. 7. Propagating type modifications to existing tables or views

The figure depicts how the type T has been replaced with T', and how an object view OV' with type T' is view of OV. The name of this relationship, IS VIEW OF, is inspired in the name of the relationship introduced in [5] to relate views with their base classes in ODMG databases.

If the external schema we are defining includes an object view OV whose type has external references, the process is the same. That is, a new type must be defined from the original type in order to hide the external references. Now, the external schema, instead of including the object view OV and its original type, will include the newly defined type and an object view OV' defined from OV. The type of OV' is the new defined type, which hides the external references of the type of the original object view OV. After, the five rules must be applied in order to propagate the changes of reduction closure.

Reduction closure algorithm

Reduction closure is carried out by an algorithm that takes as input the set of components S selected to make up the external schema. If the schema is not closed, the algorithm returns the set S incorporating the needed modifications, so that the closure of the schema is reached.

```
Procedure ReductionClosure (S)
1. List = FindComponentsWithExternalReferences (S)
2. if List ≠ NIL
3.   List = PropagateChanges (S, List)
4.   UpdateSchema (S, List)
5. end if
```

The algorithm works as follows. First, the function FindComponentsWith-ExternalReferences is called to obtain the set of components that have external references (step 1). If the schema is closed, this function returns an empty list. However, if the schema is not closed, this function returns the list of nodes corresponding to types with external references.

Once the existence of external references has been checked, if the schema is not closed, then a non-empty list is returned, and the function PropagateChanges is called (steps 2 and 3). If there exists any type affected by the replacement of types having external references with derived types, this function includes new nodes to the list. The function updates the list propagating the modifications following the rules described before, depicted in Figure 5, generating a new node for each affected type, and updating existing nodes. After propagating the changes, the list contains a node for each type that must be replaced with a new type, so that types do not have external references. All the needed information to define the new components is stored in the nodes.

Finally, when the list is built, new types and object views must be defined. Then, the schema is updated replacing with new types and object views those types, object tables and object views that have external references as well as those affected. This process is carried out by the function UpdateSchema (step 4).

4 Schema Closure in SQL:2008

Since 2008, it is available the release SQL:2008 of the SQL standard [1]. In this section we will see by means of examples the basic syntax used in SQL:2008 to define types, subtypes, typed-tables, object views and the issues to obtain the SQL code for schema closure. Nevertheless, the syntax used is the same the one proposed in SQL:1999, so that the issues described in this section can also be applied to SQL:1999.

4.1 Basic Commands in SQL:2008 to Define Object-Relational Schemas

In Section 2, we described the main features of object-relational databases, and we also mentioned that in such databases, types, typed-tables and object views may be defined. In that section we also discussed that these components can be organized in hierachies, giving rise to the existence of type hierarchies, table hierarchies and view hierarchies in the schemas. Finally, in that section we used a relationship that we called *type-of* to indicate the type of a typed-table or an object view. In order to achieve schema closure in SQL:2008, we only need to know how to define object types, object tables, object views, and how to define subtypes and subtables of existing types and tables, respectively.

To define an object type in SQL:2008, we have to indicate the name of the type, and the name and data type of the constituent pieces, which can also be object types. In SQL:2008, types are created with a CREATE TYPE statement. Once defined the object type, we can create a typed-table to hold instances of the object type. The following example illustrates the definition of the object type person-type introduced in Section 2.

```
CREATE TYPE person-type AS (
ss-number varchar(10),
first-name varchar(50),
last-name varchar(50),
age int,
address varchar(50));
```

To define a typed-table we will use a CREATE TABLE statement indicating the type of the table as described below. The example creates a typed-table person-table to hold instances of the type person-type.

```
CREATE TABLE person-table OF TYPE person-type;
```

The definition of object views is carried out adding to the view definition the type of the view as described below. The example defines a view of person-table for selecting those whose age is less than 40 years old. Given that this view makes only a selection and does not make a projection, the type of the view is the same as the type of its base table (i.e. person-type) as is shown in the code.

```
CREATE VIEW YoungerThan40-view OF TYPE person-type AS
select *
from person-table
where age < 40;
```

The definition of subtypes, subtables and subviews is carried out adding UNDER to the type, table, or view we are defining as well as the name of the supertype, supertable or superview as described below, taking into account that multiple inheritance is not allowed in SQL:2008.

The first statement of the next example creates a type student-type as a subtype of person-type adding new properties. The second statement creates a subtable of person-table, whose instances are of the type student-type defined in the first sentence. The last statement creates a subview students- Younger-Than40-view that selects those students having an age lower than 40 years old. Given that the view has been defined using a selection only, its type is the same as the type of its base table, that is, student-type, as is shown in the statement.

```
CREATE TYPE student-type UNDER person-type AS (
IQ int,
entry-date date);
CREATE  TABLE  student-table  OF  TYPE  student-type  UNDER
person-table;
CREATE  VIEW  studentsYoungerThan40-view  OF  TYPE  student-
type UNDER YoungerThan40-view AS
select *
from student-table
where age < 40;
```

The next figure illustrates the effect of these definitions in the data dictionary.

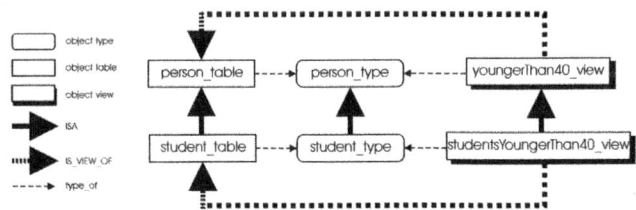

Fig. 8. Data dictionary after running the preceding definitions

4.2 The Positioning Problem in SQL:2008

Enlargement closure can be carried out in SQL:2008 just including a definition for each schema component, both those selected by schema definer and those that are referenced.

However, if we want to obtain in SQL:2008 the resulting schema of applying reduction closure, the translation is not straightforward. The problem we find is because of reduction closure may require the modification of the type hierarchy. This is a well-known problem in OODBs, named the *positioning problem*, that is related to find the right place for a type in the type hierarchy identifying existing subtype and supertype relationships with existing types. In [8], we can find some alternatives to solve this problem.

In order to illustrate the positioning problem, let us suppose that from `person-table`, whose type person-type includes the properties `first-name`, `last-name`, `age`, `address` and `ss-number`, we are interested in creating an object view, named `personWithoutAge-view`, which hides the age of people. For this view we can create a type `personWithoutAge-type` that has the same properties that `person-type` except age. This new type is a supertype of `person-type`, so that person-type must be modified to be subtype of `personWithoutAge-type`. This modification is transparent to other schema objects that depend on person- type because the type itself remains the same. The only that has changed is that some of its properties now are inherited from a more generic type.

Next, we describe how the previous example can be carried out. Figure 9.a illustrates the table `person-table` defined from the type `person-type` before creating the view `personWithoutAge-view`. Figure 9.b illustrates an excerpt of the data dictionary after defining the view `personWithoutAge-view`. As the figure illustrates, a new type `personWithoutAge-type` has been defined, and it has been integrated in the type hierarchy. Besides, the figure shows the relationship `IS-VIEW-OF` between the view and its base table to represent that `personWithout-Age-view` is a view defined from `person-table`. As can be seen in the figure, the proposed modifications do not affect to schema components defined before the modification, because `person-table` is still based on `person-type`, which still preserves its properties.

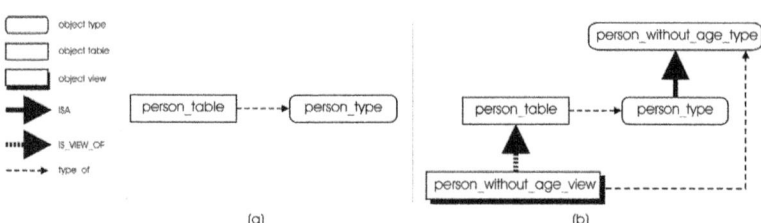

Fig. 9. a) A fragment of the data dictionaty before defining the view; b) the fragment of the data dictionary after defining the view

Therefore, applying reduction closure to a schema may entail the modification of the supertype of a type. However, SQL:2008 does not allow modifying the supertype of a type, as well as does not allow modifying types if there are schema objects that depend on the type. So, reduction closure and, in general, any operation involving the modification of the type, table or view hierarchies cannot be carried out directly in SQL:2008. A solution to overcome this limitation may be based on the next idea. If we have to change the supertype of a type (or the supertable of a table, or the super-view of a view), a new definition of the subtype has to be created specifying the new supertype (or the new supertable, or the the new superview, respectively). This may be carried out deleting first the type to be modified, and creating after the new type, preserving the name of the deleted type. The new type will have implemented the needed modifications. (That is, if A has no supertypes and now it has to be subtype of a type B, first we have to delete A and then, we have to create a new type A that is subtype of B.) However, this solution does not finish here because if the type to be modified (i.e. replaced) is referenced by other types, tables or views, the type cannot be deleted until no schema component has references to it. In such a case, before deleting the type, all the types, tables and views that are using it must stop referencing it and update their references to the type that is going to replace it. However, and as can be supposed, these types, tables and views that we have just modified can also be used by other schema components and the show goes on until the last link of the chain is reached. This link is really a schema component that is not referenced by anything. Then, in order to optimize this solution, before defining types, tables and views that have direct references to the type to be modified, we must find which are the schema components affected by the modification and start from the last links of the chain.

5 Related Work

Schema closure is a well-known problem in object-oriented databases. In such a context, schema closure involves that no class of the schema has references to classes that are not included in the schema. In this kind of databases the most used approach is enlargement closure [10,7], which proposes the recursive inclusion of referenced classes [2,7]. In [6], the inclusion of the necessary classes is proposed, but without indicating how; in addition, given that in that work external schemas only can include view classes, they propose a "virtualization" process, so that if a base class is needed to achieve schema closure, a view class is generated for this class and the view, but not the base class, is included in the schema. Regarding to the solutions proposed to reach schema closure, apart from the closure algorithm proposed by Rundensteiner in [10], the remainder of papers about schema closure does not include a detailed solution for such a problem, discussing it superficially.

If we focus in OODBs defined following the ODMG standard [3], whose object model allows the definition of interfaces and classes, an ODMG schema is closed if the next two conditions are fulfilled: on the one hand, no interface can include references to interfaces that are not included in the schema; on the other hand, no class can include references to interfaces or classes that are not included in the schema. In ODMG databases, most of the works [5,9] propose the inclusion of referenced components, that is enlargement closure, but without indicating how to achieve it. Reduction closure was introduced in [11].

6 Conclusions

In this paper, the concept of schema closure in ORDBs and the algorithms to achieve it have been proposed. Schema closure has been studied deeply in OODBs, but to the best of our knowledge it has not been studied so deeply in ORDBs. However, the closure of a schema is so important in ORDBs as in OODBs. Schema closure states that all the referenced components (object views, types, and so on) of a schema must be included into it. The most extended approach adds all the referenced components to the schema in order to fulfill this property, and we call it *enlargement closure*. In this work, an additional schema closure method, named as *reduction closure*, has been discussed. Unlike the enlargement closure, it replaces the components that have external references with view components that hide those references. In addition, this replacement must be propagated and such a propagation may entail the definition of other view components. Reduction closure carries out this propagation automatically, simplifying the definition of external schemas.

Finally, in this work we have shown how the closure concepts introduced in this paper can be applied in SQL:2008. In addition, the issues to overcome the limitation of SQL:2008 for modifying which is the supertype, supertable or superview of a type, object table or object view, respectively, or for modifying the type of an object table or object view have also been discussed.

Acknowledgements

The authors would like to thank the anonymous reviewers for their helpful comments on earlier versions of this paper.

References

1. ISO IEC 9075-*:2008. Database languages - SQL (2008)
2. Abiteboul, S., Bonner, A.J.: Objects and views. In: SIGMOD Conference, pp. 238–247 (1991)
3. Cattell, R.G.G., Barry, D.K.: The Object Data Standard: ODMG 3.0. Morgan Kaufmann, San Francisco (2000)
4. Eisenberg, A., Melton, J., Kulkarni, K.G., Michels, J.-E., Zemke, F.: SQL:2003 has been published. SIGMOD Record 33(1), 119–126 (2004)
5. Garcia, J., Ortin, M.J., Garcia, G.: Extending the ODMG standard with views. Information & Software Technology 44(3), 161–173 (2002)
6. Guerrini, G., Bertino, E., Catania, B., Molina, J.G.: A formal view of object-oriented database systems. TAPOS 3(3), 157–183 (1997)
7. Lacroix, Z., Delobel, C., Breche, P.: Object views. Networking and Information Systems 1(2-3), 231–250 (1998)
8. Motschnig-Pitrik, R.: Requirements and comparison of view mechanisms for object-oriented databases. Information Systems 21(3), 229–252 (1996)
9. Roantree, M., Kennedy, J.B., Barclay, P.J.: Providing views and closure for the object data management group object model. Information & Software Technology 41(15), 1037–1044 (1999)
10. Rundensteiner, E.A.: Multiview: A methodology for supporting multiple views in object-oriented databases. In: Proceedings of the 18th International Conference on Very Large Data Bases (VLDB), pp. 187–198 (1992)
11. Torres, M., Samos, J.: Closed external schemas in object-oriented databases. In: Mayr, H.C., Lazanský, J., Quirchmayr, G., Vogel, P. (eds.) DEXA 2001. LNCS, vol. 2113, pp. 826–835. Springer, Heidelberg (2001)

A Comparative Study of the Features and Performance of ORM Tools in a .NET Environment

Stevica Cvetković and Dragan Janković

Faculty of Electronic Engineering, Aleksandra Medvedeva 14,
18000 Niš, Serbia
{stevica.cvetkovic,dragan.jankovic}@elfak.ni.ac.rs

Abstract. Object Relational Mapping (ORM) tools are increasingly becoming important in the process of information systems development, but still their level of use is lower than expected, considering all the benefits they offer. In this paper, we have presented comparative analysis of the two most used ORM tools in .NET programming environment. The features, usage and performance of Microsoft Entity Framework and NHibernate were analyzed and compared from a software development point of view. Various query mechanisms were described and tested against conventional SQL query approach as a benchmark. The results of our experiments have shown that the widely accepted opinion that ORM introduces translation overhead to all persistence operations is not correct in the case of modern ORM tools in .NET environment. Therefore, at the end of this paper we have discussed some reasons for insufficiently wide-spread application of ORM technology.

Keywords: Object-relational mapping (ORM), persistence, performance evaluation.

1 Introduction

Modern information systems rely heavily on relational databases, mostly because of their reliability and standardized query language. Design of these kinds of systems is based on layered software architecture which implies three logical layers: presentation, business and data layer. Implementation of the data layer is essential step in the process of information system development and could take up to 40% of complete development time [5]. It is usually developed in the object oriented environment, where objects are used to represent data that needs to be persisted in a relational database. Traditional data layer development approach has shown numerous disadvantages that could be summarized in the following:

- The "object-relational impedance mismatch" problem occurs because the object-oriented and relational database paradigms are based on different principles [2]. The impedance mismatch is manifested in several specific differences: inheritance implementation, association implementation, data types, etc.

A. Dearle and R.V. Zicari (Eds.): ICOODB 2010, LNCS 6348, pp. 147–158, 2010.

- Development of advanced data layer mechanisms, such as concurrency control, caching or navigating relationships could be very complex for implementation.
- Frequent database structure changes could cause application errors due to mismatch of application domain objects with latest database structure. Those errors can't be captured at compile-time and therefore require development of complex test mechanisms.
- Developed information systems become significantly dependent on concrete database managing system (DBMS), and porting of such a system to another DBMS is complex and tedious task.

Object Relational Mapping (ORM) tools have been developed in an attempt to overcome the above listed problems. As described in the agile methodology for software and databases development [2], these tools automatically create data layer and provide a mapping between the application object model and the database relational model. They act as an intermediary between an object oriented code and a relational database. The most successful examples of ORM tools are Hibernate [3], Oracle TopLink [9] and recently introduced Microsoft Entity Framework (EF) [1], [4]. With these tools, application developer is encouraged to think in terms of data layer objects (persistent objects) and their relationships. The system takes over all the details of handling objects and relationships at runtime. It automatically tracks updates made to the objects and performs the necessary SQL insert, update, and delete statements at commit time. This way, business layer development can be done in the comfort of object-oriented languages which dramatically reduces the overall complexity and increases code understandability in applications.

However, application of ORM tools requires a very deep understanding of the concrete object model, the relational model and the DBMS used. All software design and implementation decisions should be done with this fact kept firmly in mind, since a large part of the transferring of persistent objects to the database is performed by the underlying DBMS. Different DBMS-s can implement common database functionalities (e.g. integrity control) in specific way, causing a part of the software errors to be produced by the underlying DBMS. This is the reason why the ORM generated data layers are not fully database independent.

The aim of this paper is to investigate the usage and performances of the most commonly used ORM tools in a .NET environment – NHibernate and EF. Beside a comparison of the tools from a software development point of view, we concentrate heavily on the performance analysis in order to check the common opinion that ORM tools introduce a significant slowdown compared to conventional data layer approach. In the literature there is already a comparison between different Java based ORM tools and object databases [6], [10], [11]. However, to the best of our knowledge, no scientific studies have been made comparing the different .NET ORM tools such as EF and NHibernate.

In the rest of the paper ORM specific concepts are analysed for both tools. After that, different query mechanisms are described, including query samples. Finally, performance test results are presented and discussed.

2   ORM Tools: Entity Framework vs. NHibernate

EF and NHibernate are compared from a software development point of view. Summary overview of compared features is given in Table 1. More detailed comparison is given below.

Table 1. Comparison of features between Entity Framework and NHibernate

Feature	Entity Framework	NHibernate
Programming languages	.NET (C#, VB)	.NET (C#, VB), Java
Generation of XML mapping file	Automatic	Only with 3rd party tools
Bi-directional relationships	Yes	Yes
Transactions handling	Automatic	Automatic
Locking mechanisms	Optimistic and pessimistic	Optimistic and pessimistic
Optimization	Lazy loading, caching	Lazy loading, caching
Query methods	LINQ, Entity SQL, SQL	HQL, Criteria API, SQL
Supported DBMS	MS SQL Server, MySQL, commercial for other DBMSs	MS SQL Server, Oracle, MySQL, PostgreSQl,...

2.1   Mapping Mechanism

In the context of ORM tools, Ambler [2] provides the following definition of mapping: "Mapping: The act of determining how objects and their relationships are persisted in permanent data storage, in this case a relational database". Mapping mechanism, that establishes a relationship between the persistent objects and the data stored in the database, is the backbone of an ORM tool (Figure 1). There are a number of technical challenges that have to be addressed by any mapping solution. It is relatively straightforward task to build an ORM that uses one-to-one mapping to expose each row in a database table as an object. However, when dealing with the relationships, inheritance mappings, multi-DBMS-vendor support and performance issues, mapping mechanism could become extremely complex.

Both EF and NHibernate provide a comprehensive mapping mechanism, and have GUI tools to increase developer productivity. While EF includes a collection of design-time tools for automatic mapping schema generation (integrated in Microsoft Visual Studio IDE), NHibernate mapping can be generated only by 3rd party tools. Commonly used approach is describing mapping details in XML files. The syntax and structure of mapping files are specified using a declarative language that has well-defined semantics and covers a wide range of mapping scenarios. Both tools support bidirectional relationships when the objects on both ends of the relationship contain references of each other.

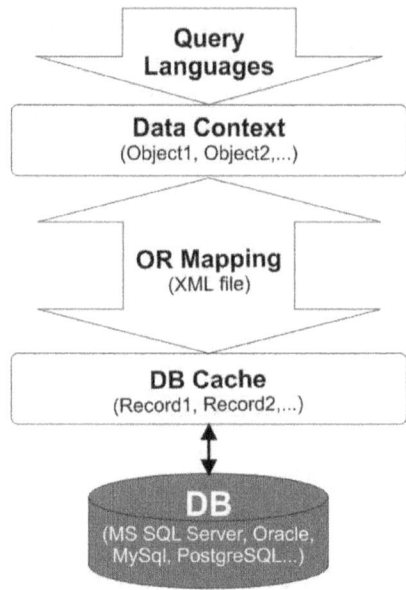

Fig. 1. Simplified general architecture of ORM Tools

Compared to NHibernate, EF introduces more general mapping mechanism represented with three separate XML sections incorporated in one file [1]. The bottom-level section of mapping file uses Store Schema Definition Language (SSDL) to describe the data source (tables, columns, constraints, etc.) that persist data for applications. Conceptual Schema Definition Language (CSDL) is used to declare and define the entities and associations of the object model being designed. The programmable data layer classes are built from this schema. The third section is written in Mapping Specification Language (MSL) which connects (maps) the declarations in the CSDL section to the data source described in the SSDL section of mapping file.

2.2 The Data Context

EF and NHibernate both provide a private cache of database persistent objects for the application execution in one thread. EF *ObjectContext* and the NHibernate *Session* are terms used for this common concept that we will call *Data Context*. The Data Context is a core which handles the loading and saving of persistent objects, where database updates are deferred until synchronization between the object cache and database is needed. It is also responsible for the management of database connections, transactions, concurrent access, etc.

In the typical scenario, when a new persistent object is created, it has no connection to the Data Context and must be explicitly introduced. In-memory changes to the object are then tracked until synchronization with database is explicitly requested. When the object is accessed again, the Data Context provides the needed data out of the cache, or if it is not found, the database is accessed. In both tools, persistent objects data obtained during the Data Context lifetime, is still there after the Data

Context is disposed. In the typical three-layer architecture, the persistent objects are filled out in the Business Layer and returned for display in the Presentation Layer, after the Data Context is disposed.

All EF persistent objects are inherited from one system super class, while in NHibernate there is no such a class. This EF approach could raise a question about code reusability because of .NET languages limitation that class can inherit only one base class. However, the problem is overcame with "partial classes" concept so that all the ORM tool provided code can be in one file and the application extended code in another.

2.3 Transactions and Concurrency

When the *Data Context* actually interacts with the database (insert or update operations), the database will open a transaction if none is currently open, and commit it after the statement. This process is known as "auto-commit". To extend a transaction lifetime to contain multiple database accesses, both ORM tools provide a transaction interface with methods for enclosing a database transaction. Actual propagation of any of the database actions is deferred until transaction commit is called. This triggers a synchronization of the *Data Context* with the database.

In order to maintain transaction isolation, databases rely on locking, which prevents concurrent access to particular data structures. Both tools provide optimistic and pessimistic concurrency models and always use the locking mechanism provided by DBMS (never lock persistent objects in memory). Optimistic locking strategy assumes that conflicting updates will cause an application exception that should be properly handled. Concurrency model could be defined for separate table columns, by setting XML mapping file, or directly in source code.

2.4 Query Approaches

Although both tools are designed to work with multiple query languages, including native SQL, this study will be focused on the two most widely used approaches – SQL Derivatives and Language Queries.

2.4.1 SQL Derivatives
SQL derivatives represent extensions of standard SQL that allows query definition on persistent objects instead of tables. They extend standard SQL in the following ways:

- Introduce native support for persistent objects within SQL queries (member accesses, relationship navigation, etc.)
- Add support for aggregation functions against objects (min, max, sum, average, etc.)
- Query results are strongly-typed .NET objects and not rows or columns.

EF EntitySQL and Hibernate Query Language (HQL) are derivatives of SQL, designed to support previously described mechanisms. Both tools use dot notation when referring child objects. While the query result of NHibernate query is a List object, in EF it is a special *ObjectQuery* object which contains methods to get List of objects.

2.4.2 Language Queries

One of the main disadvantage of SQL derivative queries, common to all string represented queries, is a software compiler inability to detect errors during compilation. Therefore, compiler cannot help the developer with compile-time checking of syntactic and semantic correctness, like it does for the rest of the program. In order to overcome the problems, both tools introduce approaches that we will identify as *Language Queries*. EF introduces LINQ [7], while NHibernate provides Criteria API for the purpose.

EF LINQ is an innovation in the programming languages that introduces query-related constructs to .NET programming languages. The query constructs are not processed by an external tool. They are rather expressions of the languages themselves. LINQ introduces nine new operators into .NET programming languages: from, join, join...into, let, where, orderby, group...by, select, and into. In addition, queries formulated using LINQ can run against various data sources such as in-memory data structures (Lists), XML documents and databases.

NHibernate offers Criteria API that uses actual .NET classes and methods to set restrictions for the query. Unlike LINQ, it doesn't introduce new operators into programming language; it only uses a new API for retrieving entities. Queries are constructed by composing API class method calls on corresponding persistent objects. This form of queries is usually called *method-based* queries. EF also provides this kind of query formulation, but it is not described in more details in this text, since LINQ has been proven as more robust method for query representation.

3 Performance Testing

In order to test EF and NHibernate performances and answer the question of whether ORM tools significantly harm overall performance, we will measure query execution speed against conventional SqlClient approach as a benchmark. Seven typical SQL test queries are defined, together with the corresponding HQL, EntitySQL and LINQ representations (Table 2). First five are SELECT queries chosen to cover various reading scenarios (joins, grouping, subqueries...), while the last two are INSERT and DELETE tests. For the first five queries, each query was executed 100 times, and the average execution time was calculated. For the last two, we calculated execution time for INSERT and DELETE of 100 records. All of the queries involve iterating through the returned results and sending back a number of objects found.

3.1 Testing Environment

Tests were executed on a PC with hardware consisted of an AMD Dual Core 2.51GHz processor, 2 GB of DDR2 RAM, 500GB hard disk (7300 rpm). Machine runs MS Windows XP Pro (SP2), with MS SQL Server 2005 installed with default settings and parameters. All tests were performed on a single machine in a computer laboratory. The software configuration that we used to run tests includes:

- MS Visual Studio 2010 (C#)
- .NET Framework 4.0 including Entity Framework
- NHibernate 2.0.1

- MS SQL Server 2005 – Standard Edition
- AdventureWorks database which is part of the SQL Server 2005 Sample Databases [8]. List of all database tables used for testing, including approximate number of records is as follows: Sales.Customer ≈ 20000, Sales.SalesOrderHeader ≈ 31500, Sales.SalesTerritory ≈ 10, Sales.SalesOrderDetail ≈ 121000.

3.2 Data Context Initialization

It is important to emphasize that in both cases, the initialization of Data Context is "expensive" operation due to one-time costs of establishing database connection and generating execution plan. Approaches for the initialization are different in EF and NHibernate. While NHibernate contains specific class methods for this purpose, EF implicitly initializes Data Context the first time an application executes a query. Explicit method call is a more flexible solution which allows a developer to implement Data Context initialization during the application initialization. It is much more acceptable solution from user's point of view, than to wait for certain period of time on first query execution when application is already started. EF can overcome lack of explicit initialization by calling some "unnecessary" query during application start just in order to initialize Data Context. The costs associated with the first-time initialization in both tools were around 3 seconds and were disregarded from the results presented in Table 2.

3.3 Test Results Analysis

A comparative evaluation of query performances between SqlClient, EF, and NHibernate approaches are given in Table 2. Additionally, graphical illustration is presented in Figure 2. Expectedly, SqlClient shows the best test results. However, query performances of the two described ORM frameworks were not significantly slower than SqlClient in the most of test cases.

The EF appears to perform better than NHibernate, except in case of Q4_Select_ GroupBy. The slow performance of EF for this query type can be explained with the non-optimal way in which group-by statements are translated into native SQL form. In fact, they will not be translated into SQL group-by statement. The EF group-by query, either EntitySQL or LINQ, is a hierarchical query that returns a sequence of groups, where each group contains the key and all the elements (records) that made up the group. It means that Entity Framework group-by is translated into simple SQL select query that fetches all the records which are further processed in the memory to get group-by results.

Performances of SQL set-based updating statement were intentionally omitted in previous discussion because they need careful consideration. From ORM tool point of view, this kind of statement cannot be expressed in its original form. Instead, ORM update query must be simulated in three successive steps including querying for the objects to be updated followed by in-memory properties update and finally submit of changes to the database. It introduces obvious performance overhead which is not dramatic in case of updating a small number of records. However, when dealing with applications that require large sets of records to be updated at once ("bulk updates"), ORM performance is hundred times worse. This is probably the weakest and the most

Table 2. A comparative evaluation of query performances using SqlClient, HQL (NHibernate), EntitySQL (Entity Framework) and LINQ (Entity Framework)

Id	SQL query string	Execution Time (ms)			
		SqlClient	HQL	Entity SQL	LINQ
Q1_Select_Simple	SELECT AccountNumber FROM Sales.Customer AS c WHERE c.ModifiedDate > '2001-12-06' AND c.CustomerType = 'S' AND c.AccountNumber LIKE 'AW000002%'	102.3	109.6	108.1	149.8
Q2_Select_Join	SELECT soh.AccountNumber FROM Sales.SalesOrderHeader soh JOIN Sales.Customer c ON soh.CustomerID = c.CustomerID JOIN Sales.SalesTerritory t ON c.TeritoryID = t.TerritoryID WHERE t.Name = 'Australia' ORDER BY soh.AccountNumber	55.3	65.2	55.7	61.7

Q3_Select_Subquery	SELECT c. AccountNumber
	FROM Sales.Customer c
	WHERE c.TerritoryID =
	(SELECT c1.TerritoryID
	  FROM Sales.Customer c1
	   WHERE c1.AccountNumber = 'AW00000021')
		21.7	22.3	15.4	37.9

Q4_Select_GroupBy	SELECT t.Name FROM Sales.Customer c JOIN Sales.SalesTerritory t ON c.TerritoryID = t.TerritoryID GROUP BY t.Name ORDER BY t.Name	0.7	0.9	21.3	22.5
Q5_Select_Top	SELECT TOP 1000 sod.CarrierTrackingNumber FROM Sales.SalesOrderDetail sod WHERE sod.UnitPrice > 123 ORDER BY sod.SalesOrderID	2.1	5.5	5.9	6.1
Q6_Insert	INSERT INTO Sales.Customer(TerritoryID, CustomerType, rowguid, ModifiedDate) VALUES (1, 'S', NEWID(), GETDATE())	48.9	63.5	62.5	62.5
Q7_Delete	DELETE FROM Sales.Customer WHERE ModifiedDate > @x	19.9	30.2	31.2	31.2

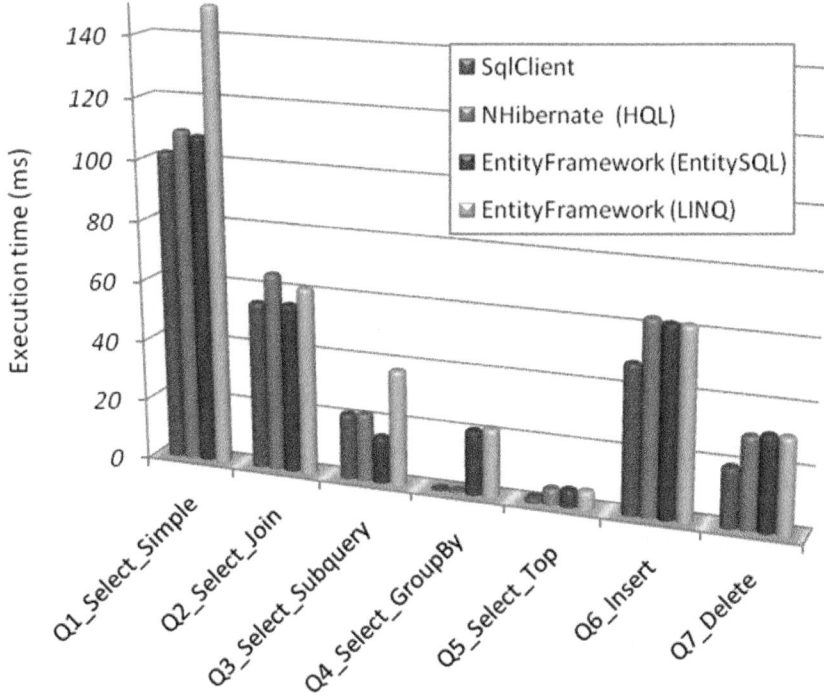

Fig. 2. Graphical representation of query performances comparison between SqlClient, HQL, EntitySQL and LINQ

criticized part of ORM concept. However, in practical applications bulk updates are not so frequent operations. In such cases there is always possibility for "hybrid" solution by explicit use of SQL updating statements via SqlClient approach.

EF offers few mechanisms for query performance improvement: use of compiled queries, defining the smart connection strings, disable change tracking, etc. According to documentation as well as some technical articles, highest LINQ performance improvement could be achieved using so-called compiled queries. It is due to obvious drawback that every time LINQ query is executed, it needs to be converted to SQL statement. The query is first checked for syntax errors by LINQ query engine and then translated into SQL statement. As a solution, compiled LINQ queries provide mechanism to cache the query plan in a static class so it can be executed much faster. Tests that were conducted on compiled and non-compiled LINQ queries have shown that the time saved with compiled queries is almost constant (in our case around 5ms) and doesn't depend on query complexity. Therefore, its relative impact on complex queries execution performance (those that last over 100ms) is insignificant. Note that all the results in Table 2 are measured for compiled LINQ queries. Also, when analyzing results, it should be taken into account that ORM tools assume additional costs of the first-time *Data Context* initialization, which were not included in presented results because they are performed only once per application.

The overall results have shown that widely held opinion that ORM is significantly slower than the conventional data layer approach for all persistent operations, is not correct in the case of modern ORM tools. Although there are some specific cases, like bulk updates or EF group-by statements, where ORM extremely degrades performances, for most of the typical data access scenarios, performances are comparable. Early ORM systems were actually slower than custom data layer solutions, because they were introducing overheads of reading metadata, reflecting on classes, generating queries, etc. In a custom data layer solution, these tasks would be completed at design time by the developer and would not affect system performance. However, modern-day ORM systems like EF and NHibernate are based on advanced architecture. They apply a variety of performance improvement mechanisms like caching, lazy loading or dirty checking that manage to compensate the overhead that ORM introduce by nature.

4   Barriers to the Adoption of ORM

Although our comparison showed that performances of modern ORM tools are close to the conventional data layer approach for the most of the typical data access scenarios, there is still skepticism toward massive adoption of ORM for information system development. In this section, we present a number of technical and cultural barriers that are responsible for insufficiently widespread application of ORM.

4.1   Learning Time Period and Costs

From the perspective of a software development company, period of time needed to learn general ORM concepts as well as specific solution details, could present serious obstacle to its adoption. Setting up and installing ORM software on all the development, testing and production machines can be a substantial undertaking. Also, new query definition approaches (like LINQ, HQL, etc.) require software developers to learn a new API that may be quite complex. Regardless of which ORM solution is chosen for adoption, it is likely that the company will have to invest in the training of software developers. However, the long-term benefits afforded by ORM can definitely outweigh these one-time personnel and technological costs.

4.2   Legacy Database Adoption

Very often, new systems need to use data from a legacy database. In this case, the software developers don't have any impact on the architecture of the database; they have to work with what is available. By the rule, legacy databases have a questionable structure reflected in poor normalization, incomplete definition of constraints, etc. With a complex OR mapping definition mechanism, it may be difficult or even impossible to represent the mapping to the legacy database. Redefinitions of OR mapping file require developers to think carefully about details that had previously been largely hidden from them. Additional problem could represent inadequate support for old versions of DBMS.

4.3 Distrust of ORM

In addition to the previously described objective barriers, there may also be bureaucratic obstacles to the adoption of new technologies, including ORM, from the ranks of project managers. Unfamiliarity with ORM concepts could represent a significant problem. Many project managers as well as software developers are not familiar with the scope of application and exact features of ORM tools. Usually, developers have outdated notions of the capabilities of ORM, believing it to be slow and incapable of supporting sophisticated database operations. Therefore, improving awareness of the features that modern ORM tools provide is an important step to their more widespread adoption.

5 Conclusion

This paper reports on the first findings of our investigation into the usage and performance of ORM technologies in .NET environment. It was found that overall performances of conventional SqlClient approach are comparable to ORM tools for the most of the typical data access scenarios. It is in contrast to the popular opinion that ORM tools add translation overhead to all persistence operations and hence are proportionally slower. ORM tools allow developers to use more powerful object oriented modeling techniques, to benefit from ease of development and provide unified programming model for different data sources. On the other side, their effective use will largely depend on the skill set possessed by project team members.

In the future, our plans are directed towards comparing ORM tools using modified versions of some standard SQL benchmarks. In addition, issues at more of an architectural level should also be investigated, for instance a comparison between distributed and a multi-user benchmark.

Acknowledgments. Work on this paper was supported in part by the Ministry of Science and Technological Development of the Republic of Serbia (Project number TR13015).

References

1. Adya, A., Blakeley, J., Melnik, S., Muralidhar, S.: Anatomy of the ADO.NET Entity Framework. In: ACM SIGMOD International Conference on Management of Data, Beijing, China, pp. 877–888 (2007)
2. Ambler, S.: Agile Database Techniques. Wiley, Chichester (2003)
3. Bauer, C., King, G.: Java Persistence with Hibernate. Manning Publications (2006)
4. Castro, P., Melnik, S., Adya, A.: ADO.NET entity framework: raising the level of abstraction in data programming. In: ACM SIGMOD International Conference on Management of Data, Beijing, China, pp. 1070–1072 (2007)
5. Keene, C.: Data Services for Next-Generation SOAs. SOA World Magazine (2004),
   http://soa.sys-con.com/node/47283
6. Kopteff, M.: The Usage and Performance of Object Databases compared with ORM tools in a Java environment. In: 1st International Conference on Objects and Databases (ICOODB 2008), Berlin, Germany (2008),
   http://soa.sys-con.com/node/47283

7. Meijer, E., Beckman, B., Bierman, G.M.: LINQ: Reconciling Objects, Relations and XML in the.NET Framework. In: ACM SIGMOD International Conference on Management of Data, Chicago, IL, USA, pp. 706–706 (2006)
8. Microsoft Download Center SQL Server, Samples and Sample Databases (2005),
    http://www.microsoft.com/downloads/
    details.aspx?familyid=e719ecf7-9f46-4312-af89-6ad8702e4e6e
9. OracleTopLink,
    http://www.oracle.com/technology/products/ias/toplink
10. Van Zyl, P., Kourie, D.G., Boake, A.: Comparing the performance of object databases and ORM tools. In: Bishop, J., Kourier, D. (eds.) Annual research conference of the South African institute of computer scientists and information technologists on IT research in developing countries (SAICSIT 2006), Somerset West, South Africa, pp. 1–11 (2006)
11. Zhang, W., Ritter, N.: The Real Benefits of Object-Relational DB-Technology for Object-Oriented Software Development. In: 18th British National Conference on Databases: Advances in Databases, Chilton, UK, pp. 89–104 (2001)

Author Index

Alagić, Suad   100

Baeza-Yates, Ricardo   6
Bernstein, Philip A.   100
Braga, Daniele   1
Büchner, Thomas   70

Ceri, Stefano   1
Chen, Tao   85
Cook, William R.   8
Corcoglioniti, Francesco   1
Cvetković, Stevica   147
Cybula, Piotr   40

de Spindler, Alexandre   55

Garví, Eladio   133
Geisler, Sandra   118
Greene, Robert   9
Grossniklaus, Michael   1, 25, 55
Guzenda, Leon   9

Jairath, Ruchi   100
Janković, Dragan   147

Keith, Michael   9
Kensche, David   118
Khan, Arif   85

Li, Xiang   118
Linskey, Patrick   9

Matthes, Florian   70

Neubauer, Peter   9
Neubert, Christian   70
Norrie, Moira C.   10, 55

Quix, Christoph   118

Samos, José   133
Schneider, Markus   85
Subieta, Kazimierz   40

Torres, Manuel   133

Vadacca, Salvatore   1
Viswanathan, Ganesh   85

Widenius, Ulf Michael   9

Zäschke, Tilmann   10
Zimmerli, Christoph   55

Batch number: 09478804

Printed by Printforce, the Netherlands